James Lenton Alty was born in 1939 three years after Turing's famous 1936 paper. After obtaining a First-Class Honours in Physics at the University of Liverpool, he obtained a PhD in Nuclear Physics which depended heavily on computer development and guided him, in 1968, to work in commercial computing with IBM (initially as a systems engineer and later as a salesman). In 1972 he was appointed Director and Professor of Computing Services at Liverpool University. It was there that he became interested in the study of the Human-Computer Interface.

Whilst at Liverpool, he was a member of the Computer Board for Universities and Research Councils and chaired an important Working Party on Microtechnology whose report – the "Alty Report" had a considerable influence on the take-up of the new technology. His team also brought the post code into common use in 1975.

In 1982 he was appointed Professor of Computer Science at Strathclyde University and was concurrently appointed the Executive Director of the Turing Institute for Artificial Intelligence based in Glasgow (1984–1989). There, he became closely involved with people who had worked with Turing. He also set up, with Heriot-Watt University, the Scottish Human Computer Interaction Centre.

Between 1990 and 2012, he was appointed Prof of Computer Science at Loughborough University. He and his research team worked with many colleagues across Europe improving the computer interfaces to large industrial plants. The team also introduced the first portable DAB radio with Roberts Radio. There is an example of their radio in the UK Science Museum.

He has been involved in many different aspects of digital computing – artificial intelligence, digital audio broadcasting, computer aided learning and interface design. In 2000, he and his research team investigated how different combinations of output media affected student learning and in 2003 produced a key paper on dyslexia.

Prof Alty is a musician and composes music. He used his musical knowledge to introduce music into the computer interface. In one well-known example, he created a Bubble Sort algorithm which output music and listeners could clearly follow what was happening by just listening to the music output!

He has experienced university life both as a researcher, and in management as a Head of Department and later as Dean of the Science Faculty. He continued to work at Loughborough University until he fully retired in 2012.

I would like to dedicate this book to my wife, Mary, who put up with my many visits abroad and kept me sane!

James Lenton Alty

IN THE STEPS OF ALAN TURING:
WORKING IN THE DIGITAL AGE

AUSTIN MACAULEY PUBLISHERS™
LONDON * CAMBRIDGE * NEW YORK * SHARJAH

Copyright © James Lenton Alty 2024

The right of James Lenton Alty to be identified as author of this work has been asserted by the author in accordance with sections 77 and 78 of the Copyright, Designs and Patents Act 1988.

All rights reserved. No part of this publication may be reproduced, stored in a retrieval system, or transmitted in any form or by any means, electronic, mechanical, photocopying, recording, or otherwise, without the prior permission of the publishers.

Any person who commits any unauthorised act in relation to this publication may be liable to criminal prosecution and civil claims for damages.

All of the events in this memoir are true to the best of author's memory. The views expressed in this memoir are solely those of the author.

A CIP catalogue record for this title is available from the British Library.

ISBN 9781398454156 (Paperback)
ISBN 9781398454163 (Hardback)
ISBN 9781398454170 (ePub e-book)

www.austinmacauley.com

First Published 2024
Austin Macauley Publishers Ltd®
1 Canada Square
Canary Wharf
London
E14 5AA

I would like to acknowledge the many people who have supported me in different ways during my career – Sir Graham Hills, Vice Chancellor of Strathclyde University, Lord Balfour of Burleigh, who was Chairman of the Turing Institute during my time as executive director, Prof Bob Churchhouse of Cardiff University and Prof Edmunds at Loughborough University. In mainland Europe, Erik Hollnagel, Rene DePlanque, Gunnar Johanssen and Peter Elzer provided much valuable advice as so did many research students who contributed during their PhD studies, in particular, Iain Duncumb, Raj Curwen, Tunu Miah, Paul Vickers and Dimitrios Rigas.

The cover image is licensed under the Creative Commons Attribution-Share Alike 2.0 Generic license. Attribution: Ian Petticrew

Table of Contents

Preface	**11**
Chapter 1: Technical Developments Leading to the Computer Age	**14**
Chapter 2: Pre-History	**21**
Chapter 3: Alan Turing's Upbringing and His Famous Paper	**26**
Chapter 4: A Baby Is Born, and War Is Declared	**30**
Chapter 5: Turing's Contribution During the Second World War	**32**
Chapter 6: The Family War Experiences	**36**
Chapter 7: Turing's Contribution Post-War; His Death In 1954	**41**
Chapter 8: Two Important Lessons Which Changed My Life	**49**
Chapter 9: Nuclear Physics Research: I Build a Small Computer	**54**
Chapter 10: At the Sharp End; Life as a Systems Engineer	**66**
Chapter 11: Computing Director, Liverpool; I Discover HCI	**87**
Chapter 12: Human Computer Interaction (HCI)	**98**
Chapter 13: Post Codes and Drama at Llanrwst	**104**
Chapter 14: Databases and Visits to Cairo	**112**
Chapter 15: The Microtechnology Revolution: The 'Alty' Report	**116**
Chapter 16: Micros Come of Age; I Go to Strathclyde University	**125**
Chapter 17 Strathclyde: Crisis at the Scottish HCI Centre	**128**
Chapter 18 The Turing Institute for Artificial Intelligence	**141**
Chapter 19 Presentations: If Things Can Go Wrong, They Will!	**154**

Chapter 20: Process Control Research: Combining AI and HCI	162
Chapter 21: Loughborough: Multi-Media and Process Control	172
Chapter 22: Composing Music	183
Chapter 23: Computing and Music	192
Chapter 24: The Digital Audio Broadcasting (DAB) Project	206
Chapter 25: The Interactive Learning Project: Dyslexia Studies	215
Chapter 26: Looking Back	227
Chapter 27: The Digital Age: The Good and the Bad	230
Bibliography	242
Appendix 1: The Felder Test	244
Appendix 2: Music on the Website	249

Preface

It was Isaac Newton who said, "If I have seen further than others, it is by standing on the shoulders of giants." One may truly say that Alan Turing was one of those giants as far as the development of computer science is concerned. His remarkable paper of 1936 laid down the basic design of the modern computer, and in 1950, his seminal paper on Artificial Intelligence set the agenda on intelligent computers for many years to come. Of course, he is now well-known for cracking the German Enigma code during the Second World War.

The book outlines Turing's contribution to computer science and his important work during the war. It also outlines my background and my life as a computing professional. From the age of 25, my life has been intrinsically intertwined with digital developments, though Alan Turing's contribution was immense compared with mine.

I was born in 1939, three years after Turing's famous 1936 paper, and became interested in computing just as computers were beginning to move into the mainstream in 1961. The early chapters discuss how digital computing developed from Charles Babbage to Alan Turing. Turing's key contribution in decoding the Enigma code during the Second World War and with Williams produced one of the earliest computers. His continued contribution after the war is outlined, particularly his paper on Artificial Intelligence. He sadly died in 1954 (possibly suicide).

After discussing two important events which changed my life, the next chapter describes life as a Nuclear Physics Research Student which depended heavily on computer development and propelled me first to work in commercial computing with IBM and later to take up academic research in computer science. I describe what it is like to work in computing at the sharp end which I hope the reader will find interesting. My last job in IBM was as a Salesperson, and that was interesting in itself!

After four years with IBM, I was appointed Professor and Director of the Computer Service at Liverpool University which involved providing computing services to whole range of users. It was there that I became interested in the Human-Computer Interface (HCI). The next chapter describes what HCI is and how it was poorly regarded at the time by many Computer Scientists.

A chapter follows describing how my research group realised the importance of the postcode and we were the first to postcode the whole of the UK and post-coded many industrial address files resulting in its use in many applications. During this work I was involved in a live TV appearance which caused something of a sensation at the time!

After joining the Computer Board, I became interested in Microtechnology and as a result was asked to form a Working Party on Microtechnology, and a whole chapter is devoted to its recommendations which had a significant effect in the UK.

In 1982, I was appointed Professor of Computer Science at Strathclyde University I first set up the Scottish HCI Centre with Heriot-Watt university, and there follows a chapter on my experiences of university politics which initially threatened to close the Centre but eventually things worked out well. I was concurrently appointed the Executive Director of the Turing Institute for Artificial Intelligence based in Glasgow (1984–1989). There, I became closely involved with people who had worked with Turing. This was interesting work and a chapter describes working with the researchers there. Chapter 19 provides a relief from computing and describes the sort of things that can go wrong (and did go wrong) when giving presentations round the world.

Working both with HCI and AI people resulted in a change of emphasis towards the problems of interfaces in Process Control of large industrial systems. Chapter 20 illustrates this work which was carried on when I became Prof of Computer Science at Loughborough University. One of my colleagues was Prof Evans who had shared a room with Turing in Manchester. Chapter 21 extends the discussion on to the use of multiple media in interfaces.

The interest in Music and my attempts to use music as a medium in HCI is discussed in Chapters 21 and 22. I hope the reader finds that interesting and do listen to the Bubble Sort!

In 1996 Digital Audio Broadcasting (DAB) had been developed in Europe but had not been seriously taken up. Chapter 23 explains what DAB is and how we became involved. It explains the implementation of DAB across Europe, the introduction of an up-channel and the production of the first portable DAB Radio with Roberts Radio. The Radio became a collector's item but the up-channel has not been taken up by broadcasters.

The final computing chapter discusses the application of multimedia techniques to Computer Aided Learning. It examines how the use of different media combinations affect learning. One interesting result of this work was its extension to examining how different the effect was on students who had Dyslexia. The effect was different than expected and resulted in a paper which was rated as the best paper at a conference in Hawaii and is still highly quoted today, 15 years later

The last two chapters reflect back on what had been achieved and an assessment of the good and poor aspects of the new technology

The aim of this book is therefore to firstly stress the importance of Turing's early contribution to the digital age, secondly to document what it was like to be a digital professional at this interesting time and some comments on the current effects of the Digital Age.

Chapter 1
Technical Developments Leading to the Computer Age

Human beings have always dreamed of creating beings in their likeness or about developing devices to ease the burdens of life. The discovery of the wheel and the capability of creating fire were huge steps in this endeavour. But it was the discovery of electricity which marked the first steps in the eventual development of the digital age. Since earliest times, man had known about the magnetic properties of lodestone and had discovered that Amber when rubbed would attract small pieces of cloth (Amber is the Greek word for "electron"). They knew that fish could give an electric shock and that there was an Arabic word for lightning. In 1600, William Gilbert wrote a treatise in which he described early experiments with "Electricus" from which our modern word "electricity" is probably derived. In 1660, Otto von Guericke successfully built a machine that could generate static electricity, and in 1752, Benjamin Franklin, by flying a kite in a dark menacing sky, received sparks on his hand.

A breakthrough, which connected electricity with living things, was demonstrated by Galvani in 1780, who showed that the signals which made muscles contract in the body were electrical in nature. He called this "Animal Electricity", and this gave a real impetus to the idea of creating an artificial human form. In 1800, Volta, working from Galvani's work, discovered that a continuous flow of electricity (an electric current) could be produced. The study of such currents was called Galvanism.

In 1816, Mary Godwin (later Mary Shelley) travelled with Percy Byshe Shelley, the poet, to Geneva to meet up with Lord Byron. It was a miserable summer (caused I understand by an excess of volcanic ash in the atmosphere) and they spent evenings together socialising and making trips to local points of interest. One evening, in a villa on the shores of Lake Geneva called Villa

Diodati, they were reading German ghost stories, and Byron challenged each person to create a ghost story. Mary Shelley (no doubt thinking of Galvani's work) outlined a story in which a Dr Frankenstein created an artificial monster with disastrous consequences. Lord Byron thought that this was a great story and encouraged Mary to develop it. Two years later, she published the now famous book, "Frankenstein, or the Modern Prometheus".

The development of computers really began in 1822 with Charles Babbage redesigning his "Difference Engine". The Difference Engine was a device for carrying out complex mathematical calculations by mechanical means. In many ways, it was like the non-electrical calculators which were common in the 1950s. It has been calculated that the first difference engine consisted of about 25,000 parts and weighed over 15 tons. To develop it, Babbage (right) received a large amount of funding.

There was, in fact, a tenuous connection between Lord Byron and Charles Babbage. In 1815, a year before the incident on Lake Geneva, a daughter had been born to Lord Byron called Augusta Ada Byron (later Ada, Lady Countess of Lovelace, usually known as Lady Lovelace). In 1833 Charles Babbage met Ada Lovelace and he showed her his "Difference Engine". Ada had very considerable mathematical abilities and they worked together for several years.

In 1837, after the failure of the first Difference Engine, Babbage embarked upon an even more ambitious project—the "Analytical Engine". The Analytical Engine was a much more complex system. Although still mechanical in nature, it had control flow in the form of branching and loops, and also had a memory for a thousand numbers, each up to 40 digits long. In fact, it had all the basic components of a modern computer (i.e., a Turing Machine) and data was to be inputted on punched cards.

Charles Babbage's Difference Machine

In 1842, Lady Lovelace wrote a comprehensive report on Babbage's Analytical Engine. In her notes, she included a detailed description of how a sequence of Bernoulli numbers (an important sequence of numbers in Mathematics) could be calculated using the Analytical Engine, and she is therefore often credited with the construction of the world's first computer program, though this is disputed, some claiming that the program was really written by Babbage. However, Babbage himself did point out that Ada found an error in one of his programs, so she is certainly credited with the discovery of the first ever "Computer Bug"!

In her notes on the Analytical Engine, she wrote that "The Analytical Engine has no pretensions whatever to originate anything. It can do whatever we know how to order it to perform. It can follow analysis; but it has no power of anticipating any analytical relations or truths". This statement of Ada was later discussed by Alan Turing in his 1950 paper on Artificial Intelligence.

The program that created the sequence of Bernoulli numbers was never run and the Analytical Engine was not built during Babbage's lifetime, since the level of Mechanical Engineering required was not available at that time, and because of a lack of funding.

Babbage was somewhat eccentric. In 1844, he contacted the poet Alfred Lord Tennyson who had recently published a poem entitled "The Wages of Sin". He wrote:

"In your otherwise beautiful poem one verse reads:

Every minute dies a man

Every minute one is born

I need hardly point out to you that this calculation would tend to keep the sum total of the world's population in a state of perpetual equipoise, whereas it is a well-known fact that the said sum is constantly on the increase. I would therefore take the liberty of suggesting that in the next edition of your excellent poem, the erroneous calculation to which I refer should be corrected as follows:

Every minute dies a man

and one and a sixteenth is born

I may add that the exact figures are 1.167 but something must, of course, be conceded to the laws of metre."

We will never know what Alfred Lord Tennyson remarked when he received this letter (but we can guess!). However, in a future edition of the poem, the lines were altered to:

Every moment dies a man

Every moment one is born.

So, Tennyson did take note!

Babbage died in 1871 without seeing the completion of either his Difference Engine or his Analytical Engine. However, in 1989 a replica of the Difference

Engine was constructed and was placed in the Science Museum in London in 2001 (the anniversary of Babbage's birth). The machine worked, proving Babbage's original design. Because of his design of the Analytical Engine, Charles Babbage is referred to as the father of the first computer. Faraday also knew of Babbage's work and held him in high regard.

Whilst Babbage and Lady Lovelace were developing the idea of a mechanical computer, A Danish Scientist, Hans Christian Oersted, discovered the connection between electricity and magnetism which was later an essential feature in the development of the electronic computer. He was demonstrating electricity to students, and, in the previous lecture, he had described magnetic properties using small magnets and compass needles (which, of course, change direction in the presence of a magnetic field). He left the compasses on the desk.

In the subsequent lecture, he turned on the electric current and, to his surprise, saw the needles change direction, showing the presence of a magnetic field caused by the electric current. This was a startling discovery which has had a huge impact on all our lives and occurred in front of a classroom of students. Later, when Oersted was congratulated on this amazing discovery, he replied, "It was just an accident", but the response was "Accidents only happen to those who deserve them!"

Shortly after Oersted's announcement, Michael Faraday in England made another astonishing discovery. He wrapped two separate coils of wire around an iron ring and found that when a varying current was passed around one coil, a current was induced in the other. This eventually led to the development of the Electric Dynamo and Transformer.

Faraday (right) made such an important contribution to our understanding of electricity that Einstein kept a photograph of him on his study wall (alongside one of Isaac Newton). Ernest Rutherford said of Faraday "When we consider the magnitude and extent of his discoveries and their influence on the progress of science and of industry, there is no honour too great to pay to the memory of Faraday, one of the greatest scientific discoverers of all time."  Faraday was keen to introduce the ideas of science to the general public and he began a series of lectures at the Royal Institution, which later became the Christmas Lectures for young children, which still continue today.

Another development took place around this time which would eventually have a huge impact on communications. This was a development using long and short electrical signals to represent numbers and letters which became known as the Morse Code.

Morse, an American (see right), had been working on the development of a system for transmitting messages electrically for several years and in 1832 the system was looking very promising. In 1835, he developed the Morse Code to enable alphabetic letters to be converted into analogous electrical signals, and in 1844 the first message "What God hath wrought" was sent from Washington to Baltimore. Western Union then developed into a major telegraph company and by 1866 had installed over 100,000 miles of wire. In fact, the telegraph was the first example of "on-line" dating! Operators were often able to deduce who the operator was at the other end of the line from the style of message construction. There were  examples of love blossoming by Morse Code, and some people even could impersonate other operators by adopting their style. There is nothing new under the sun!

Babbage died in 1873. In later life he had a passion for horse racing, and it is said that his last words before he died were "which horse won the 3.30 at Newmarket"! Both Turing's father and my grandfather were born the same year that Babbage died. As they grew up, communication between human beings was changing dramatically.

Towards the end of the nineteenth century, a message could be sent from London to New York in a matter of minutes. Just like the development of the Internet today, there were people who saw the Telegraph as a power for good, and those who saw it as a power for evil! But by 1900, just as the Telegraph was revolutionising communication, another development was about to supplant it—the telephone.

In parallel with the development of the telegraph, the telephone was being developed also using analogue signals. Like the telegraph initial development was slow. The first basic telephone was invented in 1849 and Alexander Bell (right) took out the US patent in 1875. By 1880 over 50,000 telephones  were in use, which rose to about half a million by 1900, over 2 million in 1905, and nearly 6 million in 1910. In consequence, the days of the telegraph were numbered. The telephone revolutionised person to person communication, and whilst it did not make the postal system obsolete, it added a new form of instant communication.

About this time, a further development in the use of analogue signal was being developed which would have a huge impact in the 20th Century. This was the development of machines which could record sound on disks or cylinders, which could be later played back and listened to. These inventions eventually led to the creating of the Record Industry and powerful storage devices for computers.

During this time, there was another important advance. In 1873 James Clerk Maxwell (right) formulated the classical theory of electromagnetic radiation bringing together, for the first time, electricity, magnetism, and light as manifestations of the same phenomenon. He demonstrated that electric and magnetic fields travel through space as waves moving at the speed of light.  Heinrich Hertz had already, in 1887, demonstrated transmission of electromagnetic waves from a sender to a receiver. This research eventually resulted in the development of "wireless telegraphy" or Radio, which is a key element in digital communications technology.

In the last ten years of the nineteenth century, Marconi (right) began to successfully develop radio transmission, initially across a room. There was little interest in his discovery but in 1897 he demonstrated transmission over 6 kms across Salisbury Plain and in the same year, demonstrated transmission over the sea for a distance of sixteen miles. The  message received was "Are you ready?" In 1902, he successfully transmitted a signal across the Atlantic Ocean. Marconi shared the Nobel Prize in 1909 with Karl Braun.

One of the earliest distress signals received using his system was from the Titanic. Interestingly, Marconi had originally been offered a free passage on the Titanic, but business commitments caused him to travel three days earlier on the Lusitania! By 1912, the telephone and radio telephony were providing distinct communication techniques, the former using wires and the latter by transmission through the air.

Meanwhile, another important event in the development of the modern digital computer happened towards the end of the nineteenth century. In 1875, Herman Hollerith studied mechanical engineering at Columbia University School of Mines and his PhD thesis was entitled "An Electric Tabulating System". In 1882, Hollerith (right) lectured in Engineering at MIT and there carried out early experiments on the use of punched cards. He realised from his thesis work that data could be represented by holes in these cards and the cards could be sorted in a mechanical way.

The US 1890 Census was the first Census to use Hollerith's Tabulating Machines, reducing the time taken to process the data from eight years to one year. As a result, he formed his own company (The Tabulating Machine Company) and to support the work, he developed the first card reading machine and the first Key Punch to create the punched cards.

Whilst his first machines were hard wired to process Census Data, he later developed a plug board to enable the machines to do different tasks (a sort of primitive programming). These machines were used world-wide, and he is therefore regarded as the Founder of Data Processing. His company eventually formed IBM (International Business Machines) in 1924 under the Chairmanship of Thomas J. Watson and IBM eventually became one of the most successful computer companies.

Shortly after these developments, Alan Turing came upon the scene, and it will become clear just how important a contribution he made to the development of computer technology.

Chapter 2
Pre-History

Times were tough in the latter part of the nineteenth century when my grandfather was born (1871). Both he and Turing's father were born about the same time. Whilst the landed gentry enjoyed their shooting parties and endless social occasions, life was very hard for the average family. Early death was common, and poverty was widespread.

John William Alty, my grandfather, started literally from nothing. Born in Blackburn in 1871, both his parents had died in 1876 and he became a double orphan along with his two brothers and two sisters. He was the second eldest (aged five at the time), his brother James Graham was slightly older, being seven years old and the two sisters were aged between John William and youngest boy, who was only two years old.

After the deaths of their parents, for six months, they lived in the cellars of Blackburn and begged for food, with the two elder brothers carrying the little boy on their backs. Eventually found by the police, they were taken into custody. An aunt agreed to take the elder daughter, and a farmer from the town of Haslingden (near Blackburn) called John Barnes went to a funeral in Blackburn and came back with the two boys whom he had agreed to adopt—my grandfather and his brother James.

It is a mystery as to why he adopted these boys and whose funeral did he go to? No reason was ever given—was he their father for example? After adoption, the boys kept in contact with their eldest sister but never found out what happened to the little boy, and only located their younger sister years later when she died. The farmer's wife, Alice Barnes, would never reveal anything about the origins of John William and James Graham and we always wondered why!

The boys had quite a strict upbringing. In later years, they worked on the farm early in the morning, walking the 2 miles each day to work at the cotton mill in Haslingden, then walking back to continue to work on the farm. Alice Barnes treated them quite harshly. They were never allowed to eat with the other children but ate their food sitting on the stairs. Was this because Alice knew something of their origin and events that we know nothing about?

Clough Bottom, where John Barnes's farm was located, is over the hill from a long valley called Grane, about a mile from Haslingden, which in those days was a cotton weaving town with about 30 cotton mills. The area of Grane is quite picturesque and there are three reservoirs on the left-hand side as the road rises over the hills towards Blackburn, the top one being called Calf Hey Reservoir. The original village of Grane, where my grandfather later lived, was submerged under the second reservoir—Ogden Reservoir in 1912. Prior to that, the valley was home to more than 1,300 people, which supported seven pubs, a church, and a chapel. Reminders of this lost village can be found in the ruins still partly covered by the moorland grass. The church, St Stevens, was moved stone-by-stone about a mile down the road towards Haslingden.

www.abandonedcommunities.co.uk/haslingdengrane3.html

The people of Grane were thought of as rather backward by those in Haslingden. There is one famous story about when Mafeking was relieved during the Boer War. The whole country had watched Baden Powell defend Mafeking for months and when it was finally relieved, they celebrated by taking a national holiday. One week after this holiday the people of Grane arrived in Haslingden celebrating Mafeking's release. When asked why they were there they replied "Haven't you heard? Mafeking has been relieved!" Haslingden people claim to this day that it took a week for the news to reach Grane, though the "Graners" insist that they had an important order to complete at Calf Hey Mill and so took the holiday a week later—but of course no-one believed them!

The boys were eventually thrown out of the farm (probably when they were about 18 years of age) after they took their one-week summer mill-holiday in Blackpool. Having very little money they had walked the 35-mile journey to Blackpool on the Saturday and had walked back on the following Sunday. This had enraged Mrs Barnes, who was a "good" Christian, and so she ordered them off the farm! For a time, they stayed in a lodging house in Grane Village.

John Alty vowed he would never accept being poor and during the early part of the Twentieth Century he saved every penny he could. He had two children—William Graham Alty (my father) who was born in 1900 and my father's younger sister—Annie Alty. It is said that John Alty saved money so hard that his children suffered from poor nutrition. Certainly, my father had many health problems when he was younger. For example, he was completely deaf in his left ear after a Mastoid Operation, done privately by a surgeon called Sir William Milligan probably about 1925 (which his father, who was by then quite rich, paid for privately).

John Alty eventually amassed enough money to become a sixteenth share in the building of a 1000-loom cotton mill called Grane Mill in Haslingden, and he eventually became Managing Director. The mill (see right) is still there in Haslingden, but I believe it now makes lavatory brushes! It is said that when the Union leaders argued with him over the conditions of the workers, he replied, "I have known more poverty than any of you have ever dreamed about." Apparently, when he was 40, he was given 6 months to live, but in typical fashion he responded with "You can forget that. I've too much to do."

historicengland.org.uk/listing/the-list/list-entry/1429217

The family suffered considerably in the great depression. My Grandmother (Mary Alty) used to say that my Grandfather John, would come home from Manchester and say, "Another old friend has gone bust—lost everything."

"How much did he owe you?" my grandmother would say.

"Oh about £5,000," he would reply, and put on his slippers and go to sleep by the fire! In the 30s, things were so bad that he moved the family to Blackpool and opened a chip shop. My mother used to serve behind the counter.

John Alty survived the slump and bought a car in the early 1930s. Then you did not need to pass a driving test to drive a car. Indeed 1934 was the year of the introduction of one of the most effective pieces of traffic legislation in the UK. In that year, the driving test was introduced, it was made mandatory to drive on the left (!) and Belisha Beacons at crossings were introduced (named after the Minister of Transport, Mr Hore-Belisha). That year saw the most dramatic drop in road accidents ever, and even today road deaths are much less than then.

My mother's maiden name was Beharrell which may have originated in the Netherlands. The family discovered this when they were contacted by a Sir George Beharrell (I believe he was Chairman of Dunlop) when he was trying to determine his ancestry. It turned out that the family was not closely related to Sir George, but he did give them the information he had discovered. Interestingly the name Turing might also have come from the Netherlands.

Sir George found out that a Johannes Beharrell, with his wife, Sarah Vandebek, came over to assist in the draining of the Fens in about 1650 and settled in Thornley (Cambridgeshire). In about 1840, a descendent—William Beharrell—moved to Rawtenstall in Lancashire (which is only 2 miles from Haslingden) and he married Mary Lenton. They had thirteen children, the youngest of which was Francis Beharrell (born in 1880), my maternal grandfather. In 1885, William Beharrell went by train to Peterborough to contest a will, and on his return, missed a train connection at Manchester spending the night on the station platform. As a result, he caught pneumonia and died when Francis was five years old. Francis married my grandmother, Alice Shepherd in 1909 (she was born in 1883). They had two children, Annie Beharrell (my mother) and James Lenton Beharrell (my uncle).

I do not know much about the background of Francis Beharrell. He died in 1955, and he fought in Mesopotamia in the First World War. He also had a very good bass baritone voice. My mother's family were working class and quite poor. They lived in a 2-up-2 down house in Sunnybank St. Haslingden which had no bathroom. There was only an outside toilet and one stone sink in the kitchen with a cold-water tap (the only one in the house). The outside toilet was called a "Tippler". The toilet was a hole in a wooden seat with a large drop. Halfway down the drop was a peculiar shaped receptacle which was filled by waste water from the kitchen sink. It had an odd shape and when it was nearly full it would tipple over and flush the toilet, so it was an early form of flush toilet.

Alan Turing's father, Julius Mathison Turing was born in 1873, about the same time as my grandfather. The name Turing, like my mother's maiden name, Beharrell, is also unusual and perhaps originated in the Netherlands. Julius Turing worked for the Indian Civil Service where in 1907, he met Sarah Stoney, his future wife. The Turing family, although not rich, were much better off than that of my grandfather.

Julius Mathison Turing married Sarah Stoney in 1907, and they had two sons, John, and Alan. Alan Turing was born in London on 23 June 1912, three years younger than John and roughly the same age as my mother (born in 1909). He and his elder brother John saw little of their parents because their father worked in India as a civil servant and did not return to England until 1923.

Chapter 3
Alan Turing's Upbringing and His Famous Paper

In the ten years before Alan Turing was born, huge upheavals took place in both Physics and Mathematics and in the early part of the Twentieth Century both had to be completely rethought. In 1890, it had generally been thought that very little additional information was required to complete our understanding of Physics. In that year, Max Planck (the founder of the Quantum Theory) who had just graduated in Physics, asked one of his lecturers if he could read for a PhD. in Physics. His supervisor replied that Physics was almost completely understood and that he wouldn't recommend it, since all that was left to do was a more precise determination of some of the physical constants. Fortunately, Planck, ignored this advice and still studied for a PhD. In the next ten years the traditional theories of both Physics and Mathematics fell apart. Both the Quantum Theory and the Theory of Relativity changed Physics completely, and huge changes also took place in Mathematics.

In 1922, at the age of 10, Alan (see right) was sent to a preparatory school in Sussex, and in 1926 he attended Sherbourne School, a typical English Public School. Andrew Hodges has provided a scrapbook which reveals many insights into Turing's life at that time (Hodges). He was regarded as a "difficult" boy. The headmaster wrote "If he is to stay at public school, he must aim at  becoming educated. If he is to be solely a Scientific Specialist, he is wasting his time at a public school." This very much reflects the attitude to Science and Engineering at that time indeed it still, to a lesser extent, continues today. In the UK, we use the term "Engineer" which often seems to imply someone who deals with oily engines, whereas on the continent of Europe the term Ingenieur is used which is synonymous with ingenuity. Both words are derived from the Latin.

There were indications of Turing's exceptional intelligence whilst at school, particularly in Mathematics. For example, whilst at school he read a book on Einstein's theory of Relativity and notes that he made at the time showed he understood it completely, but he was still not regarded highly at school!

In 1928, whilst Turing was still at school, a well-known mathematician, called Hilbert, disturbed by what was happening in Mathematics, had posed three questions which he felt needed to be answered:

"Is Mathematics complete?" meaning that, if complete, every mathematical statement could be proved or disproved. He assumed that there were no mathematical statements that could neither be proved nor disproved. Therefore, one should be able to show if a statement is either correct or not.

"Is Mathematics consistent?" meaning that, if consistent, an invalid statement could not be arrived at by a set of valid steps. In other words, if only proper mathematical procedures are used to create a statement, then it should be true.

"Is Mathematics Decidable?" meaning that, if decidable, there should be a general method or process by which one could determine whether a mathematical proposition could be proved or not. Given, that you have a mathematical statement, is there a method for proving it true or false?

Hilbert (right) believed that the answer to each question was "Yes". However, the first two questions were answered in the negative very shortly after Hilbert had posed them by Kurt Godel. He showed that Mathematics could neither be Consistent nor Complete.

Alan Turing entered Kings College Cambridge in 1931 as a Scholar. In his final year dissertation, he proved a famous Statistical Theorem called Central Limit Theorem (which is an important theorem in Statistics), but he subsequently discovered that it had already been proved a few years earlier. He obtained a first-class honours degree in Mathematics and was elected to a fellowship at Kings College in 1932 on the strength of his dissertation. He then studied for a PhD in Princeton University, USA between 1936 and 1938 alongside many famous mathematicians. He wrote to his mother that it was hard to be recognised when Einstein and Von Neuman were just down the hall! Turing became familiar with Gödel's theorem whilst at Princeton.

In 1936, Alan Turing wrote his famous paper "On computable numbers, with an application to the Entscheidungsproblem" which he investigated Hilbert's third question about decidability. The question can be rephrased as whether there exists a definite method—or a "mechanical process"—that can be applied to any mathematical assertion, which is guaranteed to produce a correct decision as to whether that assertion is true or not. The question however, was considered, at the time, very difficult to answer—for example—what exactly was meant by a 'method' or a 'process'?

Turing had an idea which made this method or process quite precise: — computability. At that time, "a computer" (or "Computor") was a human being carrying out a calculation (not a machine). In his paper, Turing postulated a set of procedures used by the "computer" to determine the result. The procedures were simple, being executed one at a time, but Turing showed that complex mathematical procedures could be built from these. He concluded that these procedures were so general that they would encompass anything a human being could "compute" using his method.

The idea was amazingly simple. A solvable problem could be broken down into a series of very elementary steps. Today we call this a "program". Each step pushes forward the solution of the problem. So, for example if one wanted to find the largest of, say, 10 numbers the program might look something like this:

Let the ten numbers be stored in places called A(1), A(2), A(3)……A(10).
Set Maximum =0
Take the first A(1),
Is A(1) > Maximum, if so set Maximum = A(1) If not ignore.
Now take A(2),
is A(2) > Maximum, if so set Maximum = A(2) If not ignore.
Continue this process for A(3) up to A(10)
End
At the end Maximum will have the largest value of A(1) to A(10).

Note that at each step, Maximum contains the largest number encountered so far. If a larger number is encountered, it replaces the current value of Maximum.

Although this looks like mathematics it is not the same. For example, you can have a statement like A: = A + B where the sign ": =" can be read as "becomes" rather than "equals". So, B is added to A and the result is placed back in A. As you can see, it is a sort of language (called a High-Level Language) and today there are many different High-Level languages to suit different calculation purposes. Some of you will be familiar with High Level languages such as FORTRAN, ALGOL68, PL/1 and COBOL. Note that the "computer" only carries out one instruction at a time.

In his paper, Turing also showed that you could not determine in all cases if any mathematical proposition could be proved, thus disproving Hilbert's axiom and showing that Mathematics is "Undecidable".

It was a remarkable achievement in 1936, a number of years before the first computer was built. Turing's 1936 paper is generally regarded as the most famous paper in the history of Theoretical Computer Science. Turing discovered that there are some things which cannot be computed and some of these are well-defined and understood.

Turing described his theoretical machine some years before one was built electronically. The Turing Machine is said to be universal because any computer can be implemented on another computer through software. The first "real" computers were not built until a few years later (towards the end of the 1939—1945 war). Today, if a computer can solve any problem that a Turing Machine could, using an appropriate algorithm, it is said to be Turing Complete.

Chapter 4
A Baby Is Born, and War Is Declared

It was an ordinary day in Haslingden, Lancashire on Monday, 21 August 1939, though external events were anything but ordinary. International events were moving to a climax. Hitler announced a 10-year non-aggression pact with the Soviet Union on this day causing consternation amongst other European nations. Herr Forster, the Danzig Nazi leader, on this day, addressing 4,000 of his followers at Langfuhr, near Danzig declared "Come what may, we shall see the Fuehrer in Danzig. The hour is approaching when Danzig will return to the Reich." He added: "The Fuehrer will settle Germany's problems calmly and peacefully." Also, on this day Samuel Tucker organised the first sit-in of the American Civil Rights Movement in the public library on Queen Street in Alexandria, Virginia.

A quick glance at a copy of the Times for August 21st reveals a world still having the vestiges of a bygone age. There are advertisements for cooks ("Good Cook General wanted: small modern house, North Devon, three in family, House-Parlour maid kept"); Housemaids ("Comfortable place for upper housemaid (Staffordshire), two family, six staff, own room"); and for couples ("Good post offered married couple, Parlour man and Cook, Sussex seaside, family of two, own sitting room"). There is even a German couple seeking a position as Parlour man/Housemaid in England only eleven days before the outbreak of war—I suspect they were not successful!

For about £2000 you could buy a house with 3 reception rooms, 12 bedrooms, 2 bathrooms, servants hall, with a charming but inexpensive garden.

Jobs for women were, as one might expect limited, and advertisements are quite specific about the sex of the applicant. "Wanted Young Man with radio experience"; "Numerous openings for men 18-40"; "Educated men placed in the

Motoring Industry 18-35". However, there is an advertisement for "Christian Girls (20–35) offered free training for service with Christ—Church Army"!

Haslingden is a small borough which then had a population of about 16,000 souls and is situated about 8 miles southeast of Blackburn and 19 miles north of Manchester. It is said to have the distinction of being the site of the highest Parish Church in England. In 1939 it was a reasonably prosperous mill town with over 30 cotton mills including Grane Mill.

On August 15th, the premiere of the musical "The Wizard of Oz" took place in Hollywood, starring Judy Garland as Dorothy.

On August 21st, things started well. On this day at 4pm, during a thunderstorm, a baby boy was born in Haslingden to William Graham and Annie Alty (nee Beharrell). By all account he was a good-looking baby (that's what my mother said—and she should know!). I am the smaller boy in the picture on the right. I joined my brother David Graham (born in 1933), who was seven years older than me, making a family of four.

On August 24th, the Royal Navy was put on a war footing and all remaining English Citizens in Germany were ordered home. Hitler originally intended to attack Poland on August 26th but postponed the attack for 5 days when Benito Mussolini said he could not support the invasion. On the 31st., a troop of Nazi soldiers pretending to be Poles staged a series of false operations on the Polish border to give a pretext for the coming invasion, and the next day, ten days after I was born, Hitler invaded Poland on 1 September 1939.

There are many events for which Hitler has been blamed, and one of these is the name I was given. My father, on hearing the declaration of war, immediately went to the Army Recruiting Station and joined up. As the bus was leaving for Dover, he suddenly remembered that I had not been named. Since the Registry Office was adjacent to the Recruiting Office, the Commanding Officer gave him permission to register my name before the bus left. On being asked by the Registrar what name to call me, he was initially stumped, but then responded, "Call him James Lenton after my wife's brother—she'll like that." When he arrived in Dover, he sent a telegram to my mother with the good news. She was furious—I think I was supposed to be called Michael.

Chapter 5
Turing's Contribution During the Second World War

Early, in 1919, the British Government set up the Code and Cypher Breaking Team from a group of people led by Alistair Denniston, who had pioneered code breaking during the first world war. The primary aim of the new team was to develop sophisticated code breaking mechanisms and Winston Churchill was instrumental in setting up this team. One important objective of the team was to crack a sophisticated code currently being developed by the Germans called ENIGMA. Denniston purchased an early version of a German Enigma coding machine in 1926, but the system was later considerably improved by the Germans. In 1937, the team became overloaded with messages to decode and Denniston persuaded the Government to put in more funding. He recruited several academics, mainly from Oxford and Cambridge to work in the unit. Alan Turing and Gordon Welchman were two of the Mathematicians approached and they agreed to be on an emergency list to be called up in time of war. Later in 1938, war was looming, and Turing and Welchman began working part-time at the Government Communications Headquarters in the Code Breaking Section.

In June 1938, the Head of MI6 received information that Polish Intelligence had interrogated a person who said he knew enough about the Enigma Machine to make a replica of it. He wanted money and a British Passport in return. The British Secret Service were suspicious that he might be a German Spy and Alan Turing (and another code breaker Alfred Knox) were sent to check out how much he knew. Turing and Knox were convinced that this person had an excellent knowledge of the Enigma to create a replica.

Early in 1939, Turing met two Polish Mathematicians who had constructed an Enigma Machine (The Polish Bombe) which could decode a message in 24 hours. However, for it to be useful, Turing argued that this process had to be speeded up.

As war was looming, Denniston decided that the ideal location for an enlarged code breaking centre would be at Bletchley, a town on a railway line to London and on the railway line between Oxford and Cambridge (see picture on the right). He therefore acquired the Victorian Mansion at Bletchley Park for £7,500 which also had large grounds and constructed several huts in the grounds It was called Station X. In August 1939 about 50 members of the team moved into Bletchley Park under the name of "Captain Ridley's Hunting Party"! The name of the team was shortly afterwards changed to GCHQ (Government Communications Headquarters) by which name the team is still known.

commons.wikimedia.org/wiki/
File:Bletchley_Park_Mansion.
jpg

When war was declared in September 1939, Turing moved to the code breaking centre at Bletchley Park, and he worked there for most of the war as a key member of the British Government Code and Cypher Breaking Team. The team carried out hugely important work in breaking the German Enigma Code. Turing led the team and was the main contributor to the work (along with Gordon Welshman).

Lipton tells an interesting story about Alan Turing. In 1940, worrying that England might be invaded, Turing signed up for the home guard so that he could learn how to fire a rifle. The form he had to fill in asked the question "You understand that by signing this form you could be drafted into the regular army at any moment?" Apparently, Turing signed the form but answered "No" to this question! He attended enough sessions to be able to fire a rifle, and when invasion looked less likely he gave up going.

Later he received a call to join the regular army but pointed out that he had said "No" to this question. The army checked the form and Turing was probably the only soldier who did not do 8 weeks of "Square Bashing" (i.e., marching up and down)! Of course, he would never have been drafted into the army. His work on coding was too important.

Alan Hodges tells us that Turing buried several gold bars at the onset of war but later couldn't remember where he had buried them! (Hodges)

His seminal work in cracking the German Enigma code is now well documented. The Enigma Machine (right) was an ingenious machine for encrypting and decrypting messages and the code was a very difficult code to break. This was because it depended upon what is called polyalphabetic  substitution. In a very simple code, one letter is replaced by another, but in Enigma, the substitution is continuously changing because of a series of letter rotors (you can see the rotors in the photograph) and a different start position could be chosen each day. After pressing a key on the keyboard, the character went through a series of rotors which successively altered the character. Each rotor had 26 positions (one for each letter of the alphabet). The Operator set the initial positions of the rotors and to decode the message you needed to know this initial setting. The initial machine had three rotors but at the beginning of the war, the number of rotors was increased to five. The addition of a plug board which could be set in many positions further increased the number of combinations of letters. However, the plug board had one weakness. If A was converted to say, P, then P would be converted to A.

The Polish Mathematicians who initially cracked the original Enigma code in 1932 have already been mentioned earlier. This was when the Enigma machine only had three rotors. They were able to discover the initial setting of the machine partly because the Germans transmitted the starting sequence (in code) twice before a message. The Polish Mathematicians called the decoding machine a Cryptologic Bomb (or Bombe). Alan Turing and Gordon Welchman developed a British Bombe to deal with more rotors and to cope with the fact that the Germans had stopped

From Wikimedia Commons, the free media repository

sending the starting sequence twice (this happened from the 10th of May 1940 when Germany attacked France). Turing completely redesigned the Bombe (see right) and incorporated a diagonal board (designed by Welchman) which exploited the plug-board reflection weakness, and which cut down the number of steps needed to decode a message.X

However, in 1942 the German Navy introduce the four-wheel Enigma and decoding using the bombe no longer worked, but the German Army and Airforce were still using the old enigma and Turing figured out that the machines were compatible.

In Turing's bombe a crib was used. This was a sequence of letters expected to be coded into the message (for example AND followed by a space in German) and the Bombe looked for repeated use of this crib.

Turing's code breaking techniques were very powerful and the centre was able to break the German submarine code. They were therefore able to warn convoys about the positions of submarines in the Atlantic and minimise attacks.

In parallel with the Turing's Bombe, Tommy Flowers designed the first Programmable Electronic computer at Bletchley Park called Colossus (right). It is regarded as the first Digital Computer though it was programmed by a plug-board rather than having a stored program. Turing contributed to the design though Flowers was the main designer (many people feel that Flowers did not get the full credit for the design). The first prototype went live in February 1944. Colossus provided a great deal of information to the Allies towards the end of the war, decoding messages from another coding machine called the Lorentz Machine.

Chapter 6
The Family War Experiences

My war was nowhere near as vital as Turing's work. (After all, I was only aged between 1 to 6 years old!) However, the experiences of my family during those times does shed some light into what it was like to live during those six years.

Although my father was keen to serve at the front, he was declared medically unfit for active duty (partly because of his ear problem) so he spent most of the war years at a desk in the Royal Army Ordnance Company at Burscough near Liverpool. In 1940 therefore, we all moved to Burscough where we remained until 1945 living in Trevor Road. Although I was only five years old when we moved back to Haslingden, I have only been back once to Burscough, I do have many clear memories of life at Burscough.

Children then were allowed much more freedom than they are today. I remember playing on the roofs of air raid shelters or going out with my elder brother to play in an underground den in the field. I went to school at three years old (you did then). The school was about half a mile away over two bridges. When I was four at school, I unfortunately filled my trousers in class! I tried to hide it, but the teacher quickly smelled that something was amiss and called out members of the class in turn and smelt them! When she called me out, it was obvious who was causing the smell and I was immediately sent home. It was nearly half a mile home, and I was expected to go home myself along the main road. On the way home, my mother passed me on her way to collect me and another child. She just told me to go home by myself and wait at the door. This couldn't happen, today, could it? Children were assumed to have a bit more intelligence in those days!

It was war time, but as a child it mostly passed you by. I remember the blackout, going on trains between Burscough and Haslingden in darkened carriages, and sweets on ration (only 4 ounces per week per person). I remember masses of planes in the air one day (could it have been near D Day?).

My mother's treat was, once a week, going to the cinema at the end of Trevor Road and having a coffee served at her seat. My father was supposed to babysit, but he usually arranged for two young soldiers from the camp to look after us whilst he went for a drink! My brother and I would be put to bed promptly before the soldiers arrived. As soon as they arrived, we went downstairs and we had a great time playing games until my mother was heard returning. Then we were bundled quickly into bed before she found out.

There was an aerodrome near Burscough and one incident which had a major effect on me happened in 1944. The son of a neighbour had recently obtained his spitfire wings and was anxious to show off to his parents. He flew low over the houses with his parents watching in the garden. Coincidentally we were also having tea in the garden. The pilot came in low over the house and he hit our chimney. The pilot tried to regain control of the spitfire but failed to do so and crashed in the field in front of us about 200 yards away. There were two gigantic explosions in front of us and I clearly remember black smoke with red blotches in it to this day. My father rushed out of the garden and found the pilot headless in the field. Although I do not remember any immediate adverse effects, shortly afterwards I suffered severe nightmares and had to be treated by a Psychiatrist in Liverpool. Even to this day, I have no conscious adverse memory of the incident, but it clearly affected me subconsciously at the time. Interestingly, in Spain in 2012, I met a woman (called Joy) from Burscough who knew the parents of the man who was killed. They apparently owned the pub near the bridge.

We had very few presents during the war years, but it never seemed to bother us. I remember getting a small wooden tank (clearly home-made) as a main present at Christmas and playing with it for hours. I also used to play with knitting needles on the windowsill, pretending they were trains. The other unusual incident I remember was a girl coming to school with a banana. I had never seen a banana before, and the teacher explained what it was. Frankly, it looked a bit rotten, and I decided that not only did I not like what I saw but that I did not want to see another one!

My Grandfather had offices in Manchester and used to go regularly throughout the war. One day, shortly before Christmas, he turned the corner of the street, where his offices were located, and was faced with the office having been reduced to a pile of rubble by a German bomb. He was seen running demented into the ruins lifting brick after brick. It turned out that he had a case of whisky in the office, but he never found it. He was seen shaking his fist at the sky declaring that "this time Hitler has gone too far." Ever afterwards when he told the story he claimed that this was the turning point in the war!

My Uncle Jim, the uncle I was named after, had a very eventful war. His first serious confrontation with the German army occurred at Dunkirk, in 1940, when the British Expeditionary Force was trapped and had to be evacuated. As they were retreating towards Dunkirk, my uncle and two other soldiers were given a machine gun and asked to hold a position at the end of a street. As they settled themselves down in a defensive position a large German Panzer tank rolled into the street towards them. Jim looked at his fellow soldiers, looked at the tank, looked at the machine gun which now had an uncanny resemblance to a peashooter, and said "it's time to leave", and they ran towards the beach. He was on the beach for seven days before being taken off in one of the last successful evacuations. As they crossed the channel, they were repeatedly dive bombed, but he was so tired that he slept all the way.

His second encounter with the German Army was not so successful. He was based in Crete in 1941 when the Germans launched the first paratrooper attack. Slowly the commonwealth force was pushed backwards over several mountain ranges. During the confusion, my uncle became separated from his regiment and had to fight with another regiment. This regiment was given the orders to cover the retreat by holding the last mountain pass (if necessary to the last man). Fortunately, Uncle Jim recognised a lorry from his own regiment and asked if he could re-join his regiment and he was given permission to do so. His regiment retreated to Suda Bay, and he always claimed that the other regiment at the pass did fight to the last man. On the beach at Suda Bay the troops were packed tightly together waiting for evacuation, when the Germans announced through loudspeakers that if the troops did not surrender, they would be dive bombed on the beach. Realising the hopelessness of the situation the commonwealth force surrendered, and they were marched back over the three mountain ranges. Uncle Jim was taken prisoner and spent the rest of the war in a Prisoner of War camp in Germany.

I remember my mother sending food parcels to the prison camp. He never complained about his treatment in the camp, and he received all the food parcels sent to him. He did say that food was in short supply but that the Germans suffered similarly. He once met some German prisoners of war being transferred between camps in England and he went up to them and gave them all cigarettes. I also remember him returning at the end of the war with a poisoned finger. All the flags were out at 10 Sunnybank Street (the Beharrells house) but I remember him being very quiet and not really taking part.

On his return, he married his long-time girlfriend Alice who came from Crawshawbooth in the Rossendale Valley. I remember that we had no fruit at all (apart from a few apples). Later in the war at Christmas, the Coop received a few tins of fruit. There were not enough tins to go around so there was a raffle, and we didn't get one. The next year there was a raffle again and we still did not get a tin of fruit. My mother was furious. She said that the tins of fruit in the second Christmas should have first been allocated to those who didn't get one the previous year!

We moved back to Haslingden in 1944 when V1 rockets were being fired at the UK. These were powerful rockets which contained a huge amount of explosive, powered by a powerful motor which gave off a loud throbbing sound. You could clearly hear them overhead, and when the throbbing stopped you knew they were plunging to earth. Only a few reached the north of England, but one did come over Haslingden on 24 December 1944. Everyone held their breath and eventually there was silence. It fell on Oswaldtwistle (about 8 miles away) apparently in a farmer's field.

The war with Germany ended in May 1945. I will never forget that night. We were suddenly woken up at about 2am to the sound of marching feet. I rushed to my parent's bedroom and looked out of the window (strictly forbidden in the black-out). The road was full of people marching and singing making their way to the spare land almost opposite our house. In the centre of the land was a bonfire ready for lighting (I guess they were expecting the announcement in the previous few days). We got dressed and joined everyone outside and the bonfire was lit. Everyone was so happy and joyful, and the celebrations lasted most of the night.

The victory over Japan occurred some months later with the dropping of the atomic bombs on Hiroshima and Nagasaki. I don't remember any celebrations at that time. Much of the early development work for the Atomic Bomb had been done at Liverpool University. The Professor there was Sir James Chadwick who was awarded the Nobel Prize for discovering the Neutron. Chadwick and his team then moved over to the USA for the final development of the bomb. Years later, Liverpool was still one of the main Nuclear Physics sites and it was where I eventually did my PhD.

Chapter 7
Turing's Contribution Post-War; His Death In 1954

There is a lot of argument as to who developed the first digital computer. Part of the problem is caused by the intense secrecy surrounding many of the early attempts both in the UK and on the other side of the Atlantic. Also, a Digital Computer can be defined in different ways—must it have a stored program? What sort of arithmetic can it perform? Does it store both programs and data? As already mentioned, Colossus was technically the first, though it did not have a stored program. However, it did use digital circuitry and had 1500 thermionic valves. Other contenders are EDSAC (built at the University of Cambridge), EDVAC (built in the USA) and Turing's ACE computer which eventually developed into the English Electric DEUCE machine and later the Manchester Mark 1.

Colossus has already been mentioned and was designed by M.H.A. Newman and implemented by T.H. Flowers at Bletchley Park. It was specifically designed to crack the Geheimshreiber codes (German codes specially designed to transmit highly secret information). Little was known about Colossus until the 1990's because its' design was kept secret for many years after the war. A faster version Colossus Mark II, containing 2500 valves, was produced in 1944 but it still did not operate on a stored program. Programming was achieved by altering plug boards and switches.

Maurice Wilkes, in parallel, developed the EDSAC computer (Electronic Delay Storage Automatic Calculator) in Cambridge, and this computer used Mercury Delay Line storage. Unlike the Storage tube this was not a random-access device. EDSAC is generally accepted as the first practical general-purpose stored program digital computer to be produced and it ran its first program in 1949. It had 512 words of storage in its initial form, and it served the researchers at Cambridge very effectively. It was acknowledged in the acceptance speeches of three Nobel Prize winners.

EDVAC (Electronic Discrete Variable Automatic Computer) was one of the earliest digital computers (invented by John Mauchly and J. Presper Eckert in the USA in 1944). A $1,000,000 contract for EDSAC was signed in 1946 and it became fully operational in 1949 in the Ballistics Research Laboratory.

When Colossus was decommissioned in 1945, the team who developed and worked on it (Newman, Flowers and Turing) were keen to build a stored program computer. In the same year, Alan Turing was working at the National Physical Laboratory, and in February of that year he presented the outline design of a computer he called "An Automatic Computing Engine or ACE (right) which he had partially designed whilst still at Bletchley Park. Turing deliberately included the word "Engine" in its title to pay homage to Charles Babbage who had designed the Difference and Analytical Engines. Turing's design was the first design of a stored program computer. Because of the secrecy surrounding Colossus he could not reveal that he knew it would work. Although the EDVAC computer received more publicity, ACE, in contrast to EDVAC, was more advanced in design. It implemented subroutines calls and had an early form of a Computer Programming Language.

This file is licensed under the Creative Commons Attribution 2.0 Generic license.

Turing's plan for ACE was judged to be too ambitious by his colleagues so a scaled down version—Pilot Model Ace—was implemented. It had a stored memory consisting of 12 mercury delay lines with each of these having a capacity for storing 32 words, each word consisting of 32 bits (which could take the value of 0 or 1). It was therefore truly digital, and these words could store Instructions or Data for the Instructions to work on (i.e., a program and the data words it manipulated). Since the development was rather slow (in Turing's eyes), he left before the completion of the project in 1948 and joined the University of Manchester. However, his design was completed, and it eventually ran its first program on May 10th, 1950. It was afterwards hired out to do important calculations, one of the most famous being carrying out the calculations to determine the cause of the Comet Airline disaster in 1954.

Turing joined Tom Kilburn and Freddie Williams at Manchester University in 1948. Williams had recently developed the Williams-Kilburn tube for storing digital data. This was the first random access device for storing and retrieving data. (An example of a non-random storage device would be a magnetic tape—to get at a particular datum the tape must be serially searched). A Williams-Kilburn tube could store between 1000 and 2500 binary bits, and these could be accessed randomly (this is, of course, much faster than a non-random device).

This file is licensed under the Creative Commons Attribution-Share Alike 4.0 International, 3.0 Unported, 2.5 Generic, 2.0 Generic and 1.0 Generic license.

A Williams-Kilburn tube is shown on the right. When Turing joined Williams and Kilburn the design of the machine was well advanced. The Manchester "Baby" as it was nicknamed, then developed into the Manchester Mark 1 and eventually became the first commercial computer—the Ferranti Mark 1 in 1949. Turing's main formal contribution was the writing of the first Programming Manual for the Mark 1 (with Cicely Popplewell). The first commercial version of the Mark 1 was delivered in February 1951.

In 1945, an event happened which had a pivotal effect on digital computing, but not for another 15 years. Bardeen and Shockley were awarded the Nobel Prize for the discovery of semiconductors and the invention of the transistor. At the time, the transistor was regarded as an interesting event but only of academic interest, but later, the development of the transistor changed the nature of electronics and had a huge effect on computer development.

Turing's interest in Artificial Intelligence began as early as 1947. In a report entitled "Intelligent Machinery" which he produced in 1947, he asked whether a machine could show intelligent behaviour, and in that paper, he suggested a test to check this. The test was further developed in a famous seminal paper he wrote in 1950 entitled "Computing Machinery and Intelligence". It is a very readable paper (Turing, 1950) and it is remarkable that he should have written such a paper when Digital Computers were still highly experimental.

Right at the beginning of the paper Turing realises that "thinking" is going to be difficult to define. He therefore bypasses the difficulty by redefining the question into "Can a machine do what we human beings can do?" To explore this, he invents a game called the imitation game. In the game, a man (A) and a woman (B) go into separate rooms and guests try to tell them apart by submitting questions to each room using a typewriter and examining the responses. The responses are also sent back by a typewriter so that the tone of voice cannot be used to assist in solving the problem. Player A tries to trick the interrogator, but Player B tries to help the interrogator.

Turing then replaces the man A, with a machine and asks "will the guests make as many mistakes in deducing which room contains the human being, and which contains the machine? In the standard version of the test the three participants (Computer, Human, and Human Judge) are in separate isolated rooms and the Computer and the Human both try to convince the Judge that they are human.

In the paper Turing states nine possible objections to a machine thinking:

Religious grounds: Is thinking part of the "immortal soul" which only God can create. Turing argues that "I would not be making machines with souls but rather a mansion for the souls to be in."

Head in the Sand Objections: The idea of a machine thinking is too dreadful to contemplate.

Mathematical Objections: This derives from Gödel's incompleteness theorem, meaning that a computer based upon logic cannot answer all questions. But neither can a human being!

Consciousness Argument: Can a machine write a sonnet, or a concerto based upon its inner emotional feelings. Turing replies that we have no way of telling if another human being has feelings and emotions like ourselves. (This is called the "other minds" reply)

Arguments from Various Disabilities: Can a computer be kind, fall in love, be the subject of its own thought etc. Turing claims that a machine like a human being could make a mistake, analyse its own thoughts and with enough storage capacity behave in many different ways.

Lady Lovelace Objection: Lady Lovelace insisted that a computing engine could not exhibit originality. Turing rephrased this as "a computer cannot surprise us" and argued that this wasn't true.

Continuity Argument: This claim that the brain is not Digital but Analogue in nature, but Turing argues that any Analogue system can be simulated in a Digital Computer

The Informality of Behaviour: This is a tricky one to refute. Human behaviour is not based upon well-known laws, but relies on instinct and awareness., which cannot be captured in rules. Turing argued that although we do not know the laws that govern our behaviour, this does not mean that they do not exist.

Extra Sensory Perception: ESP was quite popular at the time of the paper. He argued that conditions could be created where mind-reading did not affect the test.

Later, in the paper he examines the Lady Lovelace argument in more detail. A bright human being can be fed a few ideas and come up with a set of revolutionary deductions from them. Can a machine do this? He uses the analogy of an atomic pile. If a pile has a small mass, a neutron entering it will eventually be absorbed with little effect. But if the pile is of a critical size, one neutron could set off a nuclear reaction out of all proportion to the entry of the neutron. He therefore argues that if a computer had enough memory, the effect could be much greater than the input.

Finally, he discusses what would be needed to develop a human-like machine and likens it to child development i.e., initial state (birth), education, and experience, rather like natural selection. Even now we still do not fully understand the mechanisms by which children learn.

The paper concludes with Turing pointing out that any machine should be built using the "best sense organs that money can buy" and that chess might be a useful starting point for examining these ideas. This was an amazing paper written in 1950! Later in this book, we will see how far these ideas have progressed.

In 1951-52 Alan Turing became interested in a very different area of research—Morphogenesis. Morphogenesis is the biological process which causes a cell to develop into a particular shape. Turing was fascinated by this process—how does a biological cell know how to develop into a particular structure? One example he pondered upon was when sugar is added to hot water, it completely loses its crystalline shape, but if you cool the water the sugar eventually remembers its crystalline shape. This clearly depends upon the molecular forces that interact between sugar molecules and varies with temperature. Could similar forces have a role in Morphogenesis?

Turing suggested that a system of chemical morphogens reacting and diffusing together could become unstable and result in spatially varying patterns of chemical concentrations. These might result in patterns that we can see, for example the stripes on a zebra or a tiger's characteristic markings. Initially he developed various mathematical models of the Hydra organism, which can replace its upper structure at will. Thomson (1917) had previously predicted that the shape differences could occur because of different rates of growth in different directions. Turing successfully predicted that two competing signals—one activating growth, and another deactivating growth, would influence cell growth.

For some years, experimental biologists rejected Turing's ideas, but more recently they are being looked at again. Further development of the ideas, however, would have to wait until developments following the discovery of the structure of DNA by Watson and Crick in 1953.

He submitted his paper to the Transactions of the Royal Society at about the time he was being taken to court for being gay. Shortly afterwards, in 1954, Turing's housekeeper discovered Alan Turing's body in his bedroom. He had apparently died of Cyanide Poisoning.

An inquest was held and concluded that he had taken his own life, though his mother, Sarah Turing always maintained that he had accidentally ingested the cyanide because he was carrying out chemical experiments at that time. But if he had taken his own life, why did he do it? He had, of course, been convicted of homosexual offences the previous year and had been subjected to chemical hormone treatment over that year. However, all those who knew him were aware of his gayness and he seemed to take the prosecution quite well. There were no hints to anyone that he was about to take his own life, so we shall never know the reason he killed himself (if indeed he did that). However, whatever the reason, it left the British scientific community much poorer without him. Many years later he received a posthumous pardon from the Queen.

After Turing's death, there were developments which would change forever, the role of Digital Technology in our lives, and would result in the application and use of computers that Turing could not have possibly foreseen. Firstly, came the Transistor, which eventually led to the introduction of micro technology. The miniaturisation and the huge reduction in costs changed the role of computers in the world. In Biology, Watson and Crick's discovery of DNA changed forever our view of Biology.

In 1955, English Electric manufactured a commercial version of Alan Turing's Pilot ACE which was called DEUCE (Digital Electronic Universal Computing Engine). The computer was really fast for its time partly because Turing had insisted on a high speed of operation as the primary objective. There were three versions of DEUCE—Mark I, Mark II and Mark IIA.

Thirty-three DEUCE systems (see below right) were sold at a price of £50,000 to £60,000. These were installed in Universities (Liverpool, Belfast, Glasgow, North Staff College and The University of New South Wales, Australia) as well as in Government departments and organisations such as The Royal Aircraft Establishment at Farnborough and The Central Electricity Generating Board. Six machines were used in Aircraft design (British Aircraft Corporation (Filton and Warton), (Bristol Aircraft and Shorts, Belfast). So, Turing's initial design of the Pilot ACE had a huge impact on Commercial Computing as well as on academic Computer Science.

By 1950 the original Manchester "baby" was being further developed by Ferranti Engineers and the first system was delivered to the University of Manchester in February 1951. An improved version, the Mark 1, was then produced and seven were delivered between 1951 and 1957. The more advanced Mercury Computer was produced in 1957 and was sold successfully commercially. Ferranti sold 19 Mercury Computers. Customers included Manchester University, CERN, AERE (Harwell), the UK Met Office and the University of Buenos Aires. At this time, the UK was leading the world in computer design, but design was forging ahead in the USA.

In 1952, IBM introduced the IBM 701 computer. They claim that it was the first commercially successful computer, but in fact only 19 were sold and they mainly went to the Aircraft and Defence industries. The 1953 version of the 701 had magnetic tapes attached and the high-level scientific programming language FORTRAN (Formula Translator) was developed on the 701. The 1953 version of the IBM 701 had electrostatic storage tube memory, used magnetic tape to store information, and had binary, fixed-point, single address hardware.

In 1956, a significant upgrade to the 701 appeared, called the 704. The IBM 704 was considered an early super-computer and was the first machine to incorporate floating-point hardware. The 704 used magnetic core memory that was faster and more reliable than the magnetic drum storage found in the 701.

In 1960, IBM brought out the IBM 7090 the first to use transistors. It was the fastest computer in the world and IBM would dominate the market for the next 20 years.

Chapter 8
Two Important Lessons Which Changed My Life

As a person grows up things happen which can change that person's life, without the person realising it. There were two important events which happened as I grew up which affected my academic development and my future career, though I didn't realise their importance at the time.

I passed my eleven plus examination and originally attended Bury Grammar School for just over a year, after which my parents moved to Lytham St Annes, where I attended King Edward VII School, Lytham until going to university. Whilst at Bury I did not show any strong academic ability but was quite happy being 18th in the IIIA form. I started school at King Edwards in November 1951 and was put in the form 4A and the next year into Remove A. Unfortunately, my performance steadily deteriorated. I was about 20th in 4A and dropped to about 25th in Remove A. I do not know quite why this happened and it didn't trouble me at all. However, things went from bad to worse.

In 1954 I was put down to Lower 5X. There were three forms—Lower 5A, 5X and 5Y. 5A was the flagship form going on to take "O" levels. 5X was a mixture of hopefuls and Not-Very-Hopefuls, and the boys in 5Y were not expected to stay at school after the end of the year. Frankly, the teachers were not as dedicated in Lower 5X, and the motivation of the form was also low. One would have thought that I, coming from the A form would at least be near the top of the class. But no, I steadily drifted down the ratings until at the end of the second term I was 25th out of 27 pupils!

I went back to school in the summer term and the headmaster called me into his study. "Alty, you will be leaving the school at the end of this term since you are not bright enough to take "O" levels. What are you proposing to do after you leave?" I hadn't a clue, but I said that I wanted to be a Radio Operator in the Merchant Navy (see right!). "That's quite a good idea," he replied. What he didn't know were my reasons for choosing this career. Firstly, I thought that the uniform would attract the girls and secondly, I had the image of me sending SOS signals as the ship went down and then being classed as a hero. It never occurred to me that I might go down with the ship!

I left his study feeling quite pleased with my answer, but at the end of that week the first event occurred which change my life. I used to live about a mile from the school, and I walked to and from school with my friend Peter Smith (who was in the Lower 5A form). We walked to school across the Royal Lytham Golf Course and over some sandhills to the school (which was right by the sea). One Friday, the week after my meeting with the headmaster, we were walking home, and Peter seemed preoccupied. He actually had some notes in his hand.

"What are you doing?" I asked.

"I am learning my chemistry notes for a test on Monday," he replied.

I was gobsmacked! "What are you doing that for?" I asked.

"So, I can pass the test on Monday," he replied.

This astonished me. I never looked at any of my notes (in fact I didn't have any notes!). "What are you learning?" I asked, and I will never forget his answer.

"I am learning the Chemical Equation $Cu + 2H_2SO_4 = CuSO_4 + 2H_2O + SO_2$," he replied.

"How on earth can you learn that?" I asked.

"It's not as hard as it looks," he answered. "You see, you cannot create or destroy atoms, so if there is one S and one Cu on one side of the equation, there must be one S and one Cu on the other. Same with the O's. There are 8 on one side so there must be 8 on the other."

"My goodness," I replied, "I never knew that!"

"If you learn your notes, you'll do better in class," he said.

"But I haven't got any notes."

"Next week you can borrow these notes," he said, and I did.

The following weekend I sat down and wrote out Peter's notes and tried to learn them. One definition particularly struck me. It was the definition of Oxidation —"the increase of the proportion of non-metal in a substance". I didn't really understand it, but I learned it. The following Friday afternoon we had a Chemistry lesson. The teacher suddenly asked the class "Could anyone give me a definition of Oxidation?" I shot my hand up. This was probably the first time I had done this for years. The Teacher couldn't believe it. He said with a smile "Hold everything, we have an answer coming in from Poets' Corner!" and the class laughed. I gave my answer and it hit him like a sledgehammer. He was staggered. Because we in Lower 5X were not regarded as clever, we had been given a watered-down version of the definition. "Increasing the proportion of oxygen in a substance." My answer was, of course, the Lower 5A definition.

He did not know what to do. He explained to the class that this was a very good definition and for the rest of the lesson he kept looking at me. I honestly think he thought I had deliberately set him up. The event was a milestone in my life. I realised that revenge was sweet! It gave me a deep feeling of satisfaction and I realised that learning could have significant effects. I then applied myself to learning more of Peter's notes and, at the end of term, came top of Chemistry and 4th overall in Lower5X. I was therefore put back up into Upper 5A and allowed to sit my O levels. It was a near thing. If that had not happened, I would never have followed an academic career.

In Upper 5A I began to apply myself, but I had a long way to go. There were so many gaps in my knowledge. I decided only to take six O-level subjects— Maths, English, Physics. Chemistry Latin and French. Most other pupils took eight subjects. Shortly after I joined the class, I met another boy who caused the second event which changed my future. This pupil was Frank Duckworth. Later, Frank became well-known for inventing the Duckworth-Lewis method for managing cricket matches. Frank told me that had started a society which he called "The Philicitical Society" (I think he invented the name). To join the Society, you had to write a mathematical theorem. It turned out that there were only two members—Frank and Peter Smith! I asked to join the Society, but Frank said, "You have to submit a theorem first."

At first, I did not know what to do, but began thinking about circles. How could you work out the length of the arc of a circle given the angle subtended at the centre? For a full circle the length of the arc is $2\Pi r$ (where r is the radius and the angle at the centre is 360 degrees) so I quickly worked out that the length of the arc subtending an angle Θ degrees would have length L= (Θ/360*2*Π). I showed this to Frank, and he wasn't very impressed, but he asked, "what about the area of an arc segment?" After some thought I realised that the area of a circle was $Πr^2$ and the angle at the centre was again 360 degrees, so I guessed that the area of a segment with angle Θ would be = (Θ/360) *Π*r^2 and it worked. Frank then allowed me to join the Society.

I then thought that the same equations might work for isosceles triangles. These have two sides equal, so I reasoned that if the equal sides were of length S and the angle between them was Θ then the third side would have length L= KΘS (with a different value for K). This didn't work, and it is obvious if you draw one out, because the value of K changed with the angle Θ. I therefore made-up tables of K for different angles and I called them ISON tables denoted by Φ(Θ) so the length of the side was L =Φ(Θ)S. I eventually solved the general equation for the third side of any triangle (i.e., with three different sides) by using the fact that all triangles fit into a circle, so a triangle is actually three isosceles triangles.

Frank was so impressed he wrote the work up in a "Textbook of Isonometry" which I still have! Shortly afterwards we started doing Trigonometry at School and I realised that Isonometry was basically Trigonometry with the Φ(Θ) tables replaced by 2*Sin(Θ/2). So, it was nothing new, but it did give me a new interest in Mathematics which became important later.

I passed only 4 O levels in the summer of 1955 but passed two more at Christmas (and it was about this time that Alan Turing committed suicide). I was originally going to be a doctor, so I enrolled for Physics, Chemistry and Biology at "A" level. However, I found Biology very boring (this was before DNA) and the sex-life of the tape worm really did not interest me! After about a month a strange thing happened. Frank Duckworth, who had written the "Textbook of Isonometry" about my work in the Philicitical Society, for some reason, showed the book to the Maths Master (I do not know why he did this). The Teacher asked to see me. "Why are you not studying Maths." I replied that we had studied Calculus for one lesson a week in the previous year (It was a one lesson a week on Calculus for future Medical Students not studying advanced Mathematics) and I hadn't understood a word of it. He replied, "Well we just gave you the formulae then. We didn't really explain it. We are starting Calculus next week. Why not give up Biology and join us."

I went to the Biology Master and suggested that I gave it up. "Very wise" he replied! So, I started Advanced Mathematics and the world was spared a very poor Doctor! The first lesson of Calculus knocked me out. It was wonderful, and I understood every word of it. This was one of the best decisions I made, and I thanked Frank for his chance intervention.

I slowly improved at School and eventually obtained three A levels in Physics, Mathematics and Chemistry. I was offered a place (2nd year entry) at Liverpool University, and I started in October 1958. I had a great time at university and learned a lot about Physics and life. In 1961 I obtained a first-class degree in Physics and the Oliver Lodge prize for the best performance. I therefore decided to read for a PhD in Nuclear Physics at Liverpool.

Chapter 9
Nuclear Physics Research: I Build a Small Computer

In September 1961, I started my PhD studies at Liverpool. My research was in what is termed *Low Energy* Nuclear Physics, and in that year, I met my first digital computer—an English Electric DEUCE. Remember this computer was based upon Alan Turing's Pilot ACE computer. Indeed, my early days with computers were very much based upon machines which were derived from Turing's original designs—DEUCE, the Ferranti Mark 1, and later the English Electric KDF9 which was an advanced version derived from Turing's work.

Let me first explain what Nuclear Physics is about in non-technical terms. At the centre of an atom is a nucleus which has electrons circling round it—rather like the sun with the planets rotating around it. Nuclear Physicists want to know more about how the nucleus itself is constructed—what holds it together, for example. A nucleus cannot be seen, even under the most powerful microscope, so the only way physicists can find more about it is by firing small particles at it and seeing what happens. For example, a particle might just bounce off, or it might knock off a bit of the nucleus or it might be absorbed. So, by examining the knocked-off bits or the bounced-off bits they can tell something about the nucleus.

Because the nucleus is so small the particles fired at it must be at a high speed before the collision, and this is done using a particle accelerator. Some of you will have heard of the European accelerator at Geneva called CERN which has recently been in the news because of the discovery of what is called the Higgs Boson.

Most Nuclear Physics research involves bombarding nuclei with very high-speed particles. In 1961 there were two types of accelerator—Low Energy and High Energy. The High Energy accelerators penetrate deep into the nucleus but are not very accurate. In Low Energy Nuclear Physics, the speed of the particles is much lower, but their speed is very precise, so the measurements are very accurate. Each accelerator can yield useful but different information about the structure of a nucleus. It's a bit like fast and low speed bowlers in cricket. What happens at the wicket tells the bowler something about the batsman and how they handle the bowling.

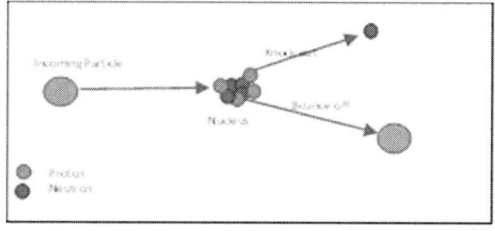

A PhD in Nuclear Physics was good training. You worked in teams, initially helping other Research Students on their research work. In the third year, more junior research students help you with your research. You received a good training in electronics, you designed apparatus, and the analysis of the results required a good knowledge of Computer Programming. It therefore trained me very effectively for a later career in Digital Computing.

When I began in October 1961, the Physics Department at Liverpool had just taken delivery of an Accelerator for Low Energy Nuclear Research, and a new building to house it had been completed next to the Chadwick Laboratory (Chadwick won the Nobel Prize for discovering the Neutron and was Prof of Physics at Liverpool for many years). The accelerator (see picture right) was supplied by a Dutch firm and several Dutch Engineers came to install it. It was a long sausage shaped device 2.5 metres in diameter and about 30 meters long. At the centre was a high voltage terminal which could be taken up to about 6 million volts. The particles were accelerated by this high voltage at the centre.

Source: Own work

To double the energy of the particles, the Tandem Van De Graaf used a trick. Before entering the accelerator, positively charged particles were made negative so that they were then accelerated up to the high voltage terminal. At the central high voltage terminal, the negative charge was instantly stripped off so that the particles accelerated again down from the high energy terminal. Therefore, you got two for the price of one! Particles with twice the energy were produced (i.e., with a 6-million-volt central terminal, particles were effectively accelerated up to 12 million volts). Of course, the inside of the accelerator necessarily had to be kept at a very low vacuum (like that in outer space).

Because the experiments were extremely accurate, the voltage at the centre had to exhibit very low variations in voltage, i.e., almost no ripple). If it varied, the energy of the particles would also vary. My first job as a PhD student was to measure the oscillations on the voltage terminal and check that they were within contractual limits. You couldn't go and measure it directly because the terminal was not directly accessible (being in the vacuum) but the accelerator had a probe which could be pushed in until it was quite near the high voltage terminal, and the other end of the probe was outside the vacuum chamber. This probe picked up only the oscillations of the terminal (not the actual voltage). It did not touch the high voltage terminal itself.

Because of the radiation emitted, no-one could be near the accelerator when it was running, and so, to measure the oscillations, I connected 500 feet of cable from the probe back to the control room, where I attached it to a Cathode Ray Oscilloscope (CRO) which shows the voltage variation graphically on a screen. So, in the control room I could measure the size of the ripple. All I had to do (when the accelerator wasn't running) was to connect a voltage oscillator of known size to the accelerator end of the coaxial cable and this allowed the CRO to be calibrated. Then I could measure the real voltage oscillations of the Van de Graaf when it was running and did this several times to see if they varied day-by-day and they were fine.

One day, I came into the control room and the accelerator was running. Someone had taken the oscilloscope (CRO) across the room to measure something else and had left my coaxial wire (from the probe) lying on the floor. I brought the oscilloscope over to reconnect it and bent down to pick up the wire. Just before I did so, the thought went through my mind—I hope no one has pushed the probe fully in so that it is touching the high voltage terminal. I dismissed this as ridiculous and touched the wire. I got the almightiest electric shock I have ever had. I actually went two or three feet in the air and as I rose up, I thought "Oh no, not 6 million volts—it's all over!" I landed on my back on the floor in a daze but actually none the worse for the experience. In fact, the probe wasn't connected to the terminal at all but because the wire had been left unconnected, but the coaxial cable had charged itself up—so it really was a huge voltage but there was very little electric charge in it, which dissipated quickly, so in the end it did no damage to me. However, I never picked up bare cables again!

I quickly had to become familiar with the huge amount of Electronic Equipment used in Nuclear Physics. There were banks and banks of electronic equipment—amplifiers, kick-sorters, gates, splitters etc. We were not really electronics experts so if we had a problem, we rang the electronics support unit (based 500 yards away at the other accelerator building). Electronics Support had a neat way of dealing with problems. When a problem occurred, the research student would ring them:

"The amplifier is not working"	"Is it plugged in?"
"OF COURSE, it's plugged in."	"Go and check it."
"I HAVE checked it."	"Is the plug switch on?"
"For goodness's sake, of course it is."	"Go and check it."
"OK, It IS on!"	"Is the fuse OK?"
"I am sure it's OK."	"Go and check it!"
"OK, the fuse is OK."	"Is the cable plugged in at the back of the box?"
"OF COURSE, it is!"	"Go and check it!"

Electronics reckoned that 80% of all faults were corrected in this procedure!

In that first year, I met my first digital computer—an English Electric DEUCE. Remember this computer was based upon Alan Turing's Pilot ACE computer. I learned to program in DEUCE Autocode (I think that is what it was called) and this was my first acquaintance with low level computer programming.

For those readers not familiar with programming, it's like writing a recipe to do some work. The computer, of course, only works with sets of "0" and "1" (called Machine Code) which are not that easy to work with, so clever people (like Turing and his co-workers) had written conversion programs which allowed mere mortals like myself to write out the recipe in a "higher level language" (a bit like English) and these programs were then converted from this higher language into machine code by a conversion program (called an Assembler or Compiler). Turing had been involved in writing one of the earliest conversion programs

So, a statement like ADD A to B might convert to something like 10111001 00011010 11110001 (the length will depend on the computer) where 10111001 is instructing the computer to ADD and the rest tells it where to find A and B. At this time, the early Assemblers were still quite low level, so the programming procedure was still quite laborious.

In the early days, the input program was usually typed onto 5-hole paper tape and the output from the computer also came out on a 5-hole tape which was then fed into a slow teleprinter to give a print-out. In those early days, at night the programmers operated the computer themselves.

Working out the physics of the nuclear reactions is quite complex and requires a lot of computing. Before I started my PhD, a theoretical physicist, in 1961, had used the DEUCE every night for six months and had been able to complete a simplified analysis for the reaction I was studying, but computing capability was moving fast and two years later, the Liverpool research group had gained access to the new Mercury computer at Manchester. We went there, using the computer during the night, and were able to do a much more complex analysis. We were able to complete a single iteration in 20 minutes, more than the theoretical physicist had done on DEUCE in 6 months! In a couple of hours, we were able to obtain a good solution. This showed how fast computers were developing.

When the particle beam came out of the end of the accelerator, it was swung through 90^0 by a large magnet and then directed down any one of three experimental lines. This enabled the three different research groups to set up the experiments, though only one group could operate at one time.

Our line ended in a large scattering chamber (about 1/2metre in diameter). In the centre of the chamber was the target and then, on the circumference was a particle counter which could be rotated through 180^0 to receive the scattered particles (See right, where the lid has been taken off). The scattering chamber was normally at a very high vacuum (because it was connected directly to the accelerator), but since it often had to be brought up to atmospheric pressure, there was a ball-valve connecting the chamber to the beam tube. To make a change to the experiment (for example change the target or counter), the accelerator was switched off. The chamber was isolated from the accelerator by closing the ball-valve, and the chamber could then be brought up to air and changes made. Then the chamber was pumped right down again. When this operation was completed, and the vacuum was very low, the ball-valve was slowly reopened to reconnect to the accelerator to the chamber.

One day I was sent to make an adjustment to the chamber. I closed the ball valve and brought it up to air pressure and made the changes. On completion of the work, I pumped the chamber down to very low pressure and then said over the intercom. "I am now opening the ball-valve" and I calmly opened it. All hell broke loose! All the vacuum gauges on our apparatus went off scale and I could hear the last-ditch emergency valves going in to save the accelerator! It seemed like I had opened the accelerator to air. I quickly closed the valve to be met by my Supervisor, Prof Green running into the experimental room, shouting "what have you done!" I thought, "Oh no, I have lost my £25 breakage deposit!"

It turned out that no damage had been done and that it was not even my fault! No one had bothered to tell me that the ball on the ball-valve had a slight scratch on it which, when it was brought up to air, a tiny amount of air was caught in it. When the ball-valve was opened to the high vacuum tube, this tiny amount of air was released and for a fraction of a second the vacuum rose momentarily to a higher (but still relatively small) level but set off all the alarms. Within a few seconds, the vacuum came back to normal. What they hadn't told me was that when the valve was opened, it should be done very slowly so that the caught air could gradually dissipate!

Whilst some people probably think that Nuclear Physicists are clever, I am afraid they make the same mistakes as anyone else! In 1963 I was President of a student hall. We had a Formal Ball every year and in 1963, the Beatles were becoming famous. Liverpool then was the centre of pop music with lots of groups playing in the city. I had heard the Beatles in their early days playing many times in the Student Union, and I had to choose a band for the Annual Hall Formal, so I went to see Brian Epstein, the Beatles manager, to get a band for the Formal. I asked him if we could have the Beatles and he said yes, but that the charge would be £50.

I replied that this price was far too high, but he responded by saying that by June, the Beatles would be internationally famous and the price then would be £120, so that £50 was a bargain. When I didn't take the bait, he offered me Gerry and the Pacemakers for £30. I replied I hadn't even heard of them, and I wasn't paying good money for unknowns. He insisted that they would also be at the top by June. Eventually, I opted for Tommy and the Challengers whom I knew well for £25. I thought I had made a bargain, but on the night of the Ball, the Beatles were top of the pops and Gerry and the Pacemakers were at number 2! So, my claims to good judgement are that I turned down the Beatles. What happened to Tommy and the Challengers? I prefer not to enquire!

When a particle is stopped by a counter the signal from the counter measures its energy. In real life, the energy of a body depends upon its speed and mass (i.e., how much it weighs). For example, if a bus and car are both traveling at 30 miles per hour, the bus has a lot more energy because of its greater weight. If both the bus and car crash into a wall, the bus will cause much more damage. One problem with nuclear experiments is that many particles can be produced by the reaction and different particles of different sizes may have the same energy (like a bus travelling at 30mph and a car at 70mph). This means the counter gives the same signal for particles with the same energy and you cannot distinguish one from the other. However, we need to know which particle is giving the signal.

To solve this problem, I had my first real encounter with computers—I built my first computer! When a bus and a car travelling with the same energy are stopped by hitting something, the energy released is the same, but if they go through a wall first and it doesn't completely stop them, the bus damages the wall more than the car. We wondered if the same could happen with nuclear particles.

We had identified some earlier research in which a very thin counter (called a DE counter, like the wall) was placed in front of the main counter (called an E counter, like the house). The incoming particle would lose some energy (DE) in passing completely through the DE counter and then be completely stopped by the main E counter. The total energy of a particle would be the sum of the E + DE counters.

Interestingly, at the time, the other researchers had observed that the signal from the particle passing through the DE counter was very different (like the bus and car) for different particles of the same energy. Imagine a car and a bus travelling through the thin wall with the same energy (the car would be going much faster because it is lighter) and then both are stopped by hitting a house. All the energy is absorbed by the wall (DE) and the house (E) and the total damage is the same. However, the damage to the thin wall will be different for the bus than for the car.

The workers had found that if you multiplied the E signal by the DE signal you could obtain a signal which was constant depending only on which particle was passing into the counters. The true formula was Signal = $(E + E_0 + \eta DE) \times DE$ where E_0 and η are constants which you can adjust depending on what energy range is being examined. If you set up E_0 and η correctly, the signal has a fixed value for one particle and a different fixed value for heavier particles (see diagram opposite).

Today, you can buy a small computer to do this, but then computers were very expensive (and large) so this was my first encounter with computers and had to build a simple analogue computer myself to achieve this. I used what are called squaring tubes.

A squaring tube is a valve which squares the input. So, if you input I it will output I^2. They are obsolete today, but then they were used quite a lot.

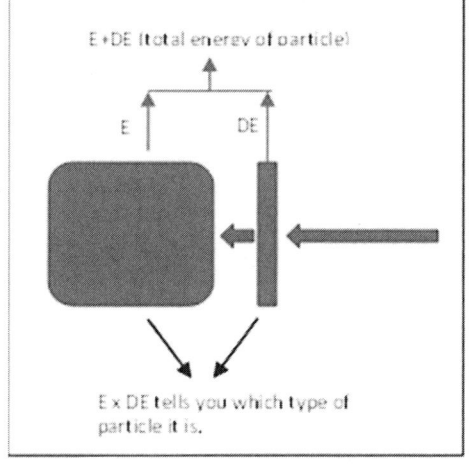

These were electronic valves (see right) which if the input is I, the analogue output is I^2.

But how can we get (a x b) by using squaring tubes? To achieve the multiplication a x b we used the algebraic relationship that

$(a + b)^2 = a^2 + 2ab + b^2$
and $(a-b)^2 = a^2 - 2ab + b^2$.
So $(a+b)^2 - (a-b)^2 = 4ab$. (Check it, it really is true!).

So, two signals needed to be created $(E + E_0 + \eta DE)$ (i.e., a above) and at the same time create the signal DE (i.e., b above). Then add them and square them $(a + b)^2$, subtract them and square them $(a-b)^2$ and take the two results away. The result will be $(E + E_0 + \eta DE) \times DE$. Read it again slowly and it's obvious!

With the right settings of E_0 and η, each particle gave a unique signal. So, I built the box in electronics, which was fascinating because I had never built any electronics before.

This simple analogue computer worked well, and it was able to direct the different incoming particles into different Kick-Sorters (a Kick-Sorter measures particle energy graphically) so they didn't interfere with each other. Note that it is not a digital computer because all the signals are continuous (i.e., analogue). This was the first computer I had built, and I was quite proud of it. Of course, only a couple of years later, you could buy a small computer to do this!

In 1965, Liverpool University had upgraded its computer to an English Electric KDF9. The KDF9 was a solid-state based machine based upon diode transistor logic. It was one of the earliest Timesharing Systems and could run up to four programs at once. It was very fast for its time being about one sixth the speed of the ATLAS computer. It is difficult to be precise about the speed comparisons, but our programs ran about 7 times faster on KDF9 than on the Mercury. Thus, on the KDF9 we could do much more extensive calculations. Towards the end of my thesis, we also ran the programmes on the ATLAS computer at Oxford. This gave us a further 8 times improvement on computing speed and we could complete an analysis in 19 seconds!

You may think that famous nuclear physicists knew all about the physics of simple objects. At this time there were two famous physicists, Erwin Schrodinger (left) and Werner Heisenberg (right)). A friend of mine once met them both at University College Dublin, where I think Schrodinger was the Prof of Physics. One evening at a college dinner,   Heisenberg was visiting Schrodinger, and they were discussing physics over dinner, when a retired English lecturer who lived in the college, suddenly turned to them and said, "Aren't you two guys physicists? Perhaps you could help me with a problem. If you take a bicycle, place the pedals in the vertical position and attach a rope to the lower pedal, and then you stand behind the bike and pull the rope from behind, does the bicycle go forward or backwards!"

Schrodinger initially said, "It will obviously go backwards because the overall force is backwards."

Heisenberg disagreed. "I am not absolutely sure (remember he was the author of the uncertainty principle!), but it will depend upon the friction of the back tyre. If it has a good grip, the bike will go forwards." They argued for about an hour and covered paper napkins with equations. My friend had to leave for another engagement, and when he returned at about 10 pm, he found Schrodinger and Heisenberg at the college gates with a bicycle and a piece of rope!

Heisenberg was famous for his uncertainty principle. Put simply, if you know exactly where you are, you have no idea how fast you are going. Alternatively, if you know exactly how fast you are going, you do not know where you are! Schrodinger is, of course, famous for his cat. In the early days of Quantum Mechanics, there was an important principle called Quantum Superposition. An atom could simultaneously exist as a combination of states corresponding to different possible outcomes. The prevailing theory said that a quantum system remained in this superposition until it interacted with, or was observed by, the external world, at which time it goes into one of the possible definite states.

To illustrate this peculiar principle, Schrodinger invented a story about a cat. Schrodinger's cat is inside a black box (not observable) and a random event in the box causes it to be poisoned and it dies, but the random event can take place in a few minutes or even after several days or years. So, until the box is opened, the cat can be alive and dead at the same time! Once the box is opened (observed), the cat will be seen either as dead or alive.

There is a famous physics joke about Schrodinger and Heisenberg. They were both travelling in a car at more than the speed limit and a policeman stops the car. Heisenberg winds down the window. The policeman says, "Do you realise that you were exceeding the speed limit?"

Heisenberg replies, "I am not exactly certain (!), I have no idea how fast we were going but I know exactly where I am." The policeman thinks that these guys are a bit odd, so he goes around to the car boot and opens it. There is a closed black box in the boot.

The policeman opens the box and exclaims, "There's a dead cat in this box!" and Schrodinger replies, "Now there is!" Years later, I told this joke at my daughter's wedding and strangely, nobody laughed, but physicists think it's a wonderful joke!

We completed the experimental runs for my thesis early in January 1965 and after the last run I went and got married to Mary (a charming Welsh girl!). For the next six months, the data was analysed using KDF9 and the ATLAS computer at Oxford. During October 1965 to January 1966, I wrote up the PhD thesis. It was much harder in those days before word processors. The student had to give each page for typing to a secretary and she would type in on a Gestetner plastic sheet which could be used on a duplicator to produce duplicate pages. If a mistake was made, it could be pasted over, but the re-typing never quite fitted neatly into the gap, so retyping the whole page was common. There were real problems if an extra paragraph was needed after typing. The words had to be juggled until the new paragraph fitted on the page without the total number of words overrunning the page. It was difficult to add extra words because this could be very costly.

The good thing about this system was that it prevented students from being too verbose! My final thesis had 100 typed pages (about 230 words per page) and 67 diagrams. Today, PhD theses are often over 300 pages long with 440 words per page (far too long!). My Viva-Voce examination took place in April with Prof Ken Allen of Oxford as the external examiner I passed with no changes, and I received the Ph D. degree at Liverpool in July 1966.

I remember a final story about Schrodinger which I personally witnessed. The Prof of Theoretical Physics at Liverpool, Prof Herbert Frohlich, was famous and it was often said that he should have received the Nobel Prize for his work on transistors. He ran the Theoretical Physics department and his second in command was a Senior Lecturer called Dr Huby. Frohlich felt very strongly that no-one should need slides in a presentation. He thought that the blackboard and chalk were more than enough to get a point across.

It was announced that Prof Schrodinger was coming to give a Seminar and the rumour spread round the department Schrodinger was bringing slides! The seminar was packed as people waited breathlessly as to what would happen when Schrodinger asked for the first slide. He did so, and a gasp went up, then Dr Huby produced one of those simple single slide viewers which he had bought at the local store and Schrodinger's first slide was solemnly passed around. The rest were past round in the same way. The seminar took about two hours!

I am afraid my thesis was not as important as Alan Turing's PhD. thesis, where he proved an important theorem in Statistics—the Central Limit Theorem. He then realised that it had been proved by someone else one year before, but this does not detract from a very considerable achievement.

I realised that there were few job opportunities in Nuclear Physics, and I had enjoyed my (limited) experience of computers. In 1966 therefore, I took a Research Fellowship for two years in the Department of Metallurgy. I worked with a new Professor there called John Stringer, who was very supportive. A number of my friends, including Frank Duckworth were in the department. After I had been there about 15 months a lectureship became available, and I asked Prof Stringer if I should apply. He replied that the Lectureship had already been promised to another person by the Head of Department (another Professor) but I should still apply showing that I was interested.

The lectureship interview happened in June. I attended the interview and found out that there were only two applicants selected for interview—myself and a Dr Bacon. I assumed he was the one to whom the job was promised and that he would be offered the post. I therefore went for interview assuming I would not get the job. Universities are not good at keeping secrets so at Lunchtime the news filtered out that the appointing committee could not agree as to which candidate should be appointed! The two professors were therefore asked by the committee to decide. At 3pm further news leaked out that they could not agree, and the Head of Department would need to make the decision himself.

At 4pm I was asked to see the Head of Department. I walked in assuming that I would not be appointed and was quite happy to accept Dr Bacon. The Head of Department explained that he had offered the job to Dr Bacon and therefore he felt he had to offer it to him. He said it had been one the most difficult decisions of his career. I then replied with what I thought was an inspired response "Don't worry about that. Now offer me one!" He replied saying that he could morally not do that to which I replied, "What a great time to be moral!"

That moment I turned from being quite relaxed to being infuriated and I said, "In that case I shall be leaving the department." He tried to persuade me to stay but my mind was made up. Shortly afterward I met Prof Cassels who said, "May I be the first to congratulate you on becoming a Lecturer in Metallurgy." I replied, "I am afraid not—they turned me down!" I had not put Prof Cassels on my list of referees in my application. So why was he asked? I suspected this was because they thought that Prof Cassels and I didn't get on (which wasn't true).

In retrospect I didn't blame the Head for appointing Dr Bacon. He was a first-class Metallurgist and was a much better appointment than me. The Head should just have told me that he was the best candidate. I do not think I would have made a good Metallurgist, so the result was beneficial both for the Department and myself!

Chapter 10
At the Sharp End;
Life as a Systems Engineer

After deciding to resign from Metallurgy, I decided to look for a career in computing. During the 1960s, commercial computers had developed at a huge pace. Computer store was becoming cheaper, and computers ran faster and were therefore able to solve much more complex problems. This huge increase in capability reflected what is called Moore's Law—the main characteristics of the modern computer—available storage, the disk space and the speed of computers were doubling roughly every 10 months! However, in 1963 the computers in the marketplace were still highly specialised. They either solved Scientific problems or Commercial problems. This all changed in 1964 when IBM announced the System 360 family of computers.

In System 360 there was no distinction between Scientific Computers and Commercial Computers—hence the name—they covered 360 degrees of the compass! Previously all computers were unique. If you wanted to increase your capability you had to buy a new machine. In System 360 a totally compatible set of models were produced from the tiny 360/20 up to powerful computers such as the 360/65 and the 360/195. Users could move up the scale as their computing became more complex, transferring their programs from one system to another. The design was hugely successful and resulted in one of the most successful computer ranges in history, influencing computer design for years to come.

Another important development was microcode technology. This was originally proposed by Maurice Wilkes from Cambridge University. With microcode, any computer (even that of different computer manufacturer) could be simulated on another computer. This is called Emulation. This feature enabled IBM to "emulate" the earlier IBM 1400 computer on a System 360, allowing customers to transfer easily to System 360 from older computers. It also enabled

IBM to implement a compatible range of computers with widely differing performance.

System 360 was very successful because customers could purchase a small system and knew that they would always be able to migrate upward to a larger computer when needed without reprogramming the application software or replacing peripheral devices.

In January 1968 I applied to IBM to become a Systems Engineer. Systems Engineers are responsible for the software of the computer system. They monitor the running of the Operating System, advise the customer on new ways of exploiting the system and regularly update the operating system. They also assist the Salespeople in trying to sell new systems or upgrade existing ones. It is a very rewarding job. I had a tough interview and had to pass their Aptitude Test, but they accepted me (Yes, I was surprised as well!). They offered me a salary of £1950 per annum. It looks small by modern standards, but it was equivalent then to that of a new lecturer at a university.

Just before I joined IBM, I was nearly killed in a climbing accident! At the end of March, I left Liverpool University and took a week's holiday climbing in Scotland with friends. I was not a real expert but could climb standard snow and ice routes. We all went up to Torridon in the north of Scotland and stayed in a small cottage right under the impressive mountain of Liathach. Liathach (right) is a big mountain and has a long (4 miles) knife-edge ridge (like Crib Goch in Wales). In  the centre of the ridge are the Fasarinen Pinnacles, which are tricky, particularly in winter.

The ascent was straightforward, gradually getting steeper and the snow was deep, and we finally reach the ridge. In winter, ridges are normally covered with a cornice of snow on the leeward side. Cornices form on the sheltered side of a ridge, and they can be 20 or 30 feet high and 4 or 5 feet deep in snow. It was decided that I would lead up the cornice (i.e., cutting steps with my ice axe and being roped for safety). As a precaution, I had two ropes attached, belayed part way up the cornice.

Climbing up the cornice was straightforward, and I reached the top to see a breath-taking view. My chest was level with the top of the ridge, and I could see the snow-covered ridge stretching away to the left. There was also a great view of the mountains of Canisp and Suilven across to the right. I called for the rest to follow, when suddenly the whole cornice gave way (presumably under my weight). I fell backwards (with the snow) and landed on my back about 60 feet below. Luckily the snow was deep and cushioned the impact. Also, the ropes partially broke my fall. I was lucky because if there had been any exposed rocks I could easily have been killed. My friends picked me up, and apart from some bruising I was OK. Since the snow had come off the ridge it was now easy to climb so we carried on and did the whole ridge. However, I decided that day—no more snow climbing—I had a wife and two children, and the risks were not worth it!

I began working for IBM in April 1968 and attended my first course as a Trainee Systems Engineer. All IBM courses were in-house run by the London IBM Education Centre. There were about 25 students on the course with very varied backgrounds. The course lasted two weeks and covered 360 Assembler (the lowest level of Machine code programming i.e., "0" s and "1" s.) and IBM Operating Systems (the systems which run the computer). Training for a Trainee Systems Engineer typically lasted about 18 months.

I should briefly explain that a computer runs under the control of an Operating System. This decides which program should run next, connects programs to printers, reads in punched cards or paper tape input, and connects with other devices. Small computers have relatively simple operating systems, but for large ones, the operation system is complex.

Over the next 18 months there would be 4 x 2-week training courses in London., and between courses a Trainee Systems Engineer was assigned to an account to gain experience. I was assigned to the Nuclear Accelerator at Daresbury, 16 miles from Liverpool. This was a forerunner of the Hadron Collider at CERN. It was a circular accelerator ring (probably about 100 metres in diameter) which could accelerate particles up to huge energies. No doubt IBM thought this was an ideal match, though my Low Energy Nuclear Physics had little in common with the High Energy Nuclear Physics that they did at Daresbury. However, the fact that I had a PhD put me, I guess, at the same academic level as the very talented workers at Daresbury. The village of Daresbury is an interesting place. Lewis Carroll was born there in 1811 and eventually wrote the book "Alice in Wonderland". Interestingly, he was also a mathematician.

At your training account you were supervised by an experienced Systems Engineer. I was very fortunate because my tutor was Tim Lloyd, and he was a great guy. He trained me well. One thing that really impressed me with IBM was the way everyone helped each other. There was no one-upmanship. If you asked for help, it was given generously.

Computers store programs and data in what are called "Bits and Bytes". A Bit is a "0" or "1". A collection of 8 Bits is called a "Byte", and a Byte can contain a character (like "A", "B" or "3") so the word "hand" would be contained in 4 bytes. Daresbury was a very interesting IBM Installation. It ran a 1 Megabyte 360/65 system, one of the finest computers that IBM produced. A Megabyte means that the computer store was over one million bytes. A Megabyte could store about a 1,000,000 letters or about 120,000 written words (or like a book of, say, 60 pages). The computer ran under the highest level of the Operating System—MVT (Multiprogramming with a Variable number of Tasks). At that time, Daresbury was the only customer in Europe running MVT. It was therefore a very important account. If the machine was down for more than 12 hours, the UK Chairman of IBM was informed!

MVT is a complex operating system and took me some time to understand it. Our job was to keep the system running, identify faults, correct them and encourage Daresbury to use more advanced facilities. The staff at Daresbury were also very competent. Trevor Daniels was the Systems Manager, and they had some very good programmers. I spent four months there before I was sent on my second training course.

At the end of each two-week IBM course, students had to sit a Multiple-Choice exam on what they had learned. When I joined IBM, I thought that I (having a PhD) would breeze through the exams. However, the multiple-choice exam didn't suit me too well. Although I understood the questions, I would often make a slight mistake and end up with the wrong answer. I obtained reasonable marks but was certainly not top of the class. I assumed this wasn't a problem. I knew what I was doing—it was just a slight error.

When I returned to the Liverpool Office, my manager asked me why I hadn't done quite as well as he had expected. I replied confidently that I had understood the questions but that I had made slight mistakes which resulted in me getting the wrong answer. However, I said, brightly, surely understanding mattered more than making occasional slips. This was a serious academic error! His face was crestfallen! He pointed out that I was not in academia now. What matters in IBM is getting it right—we don't care if your reasoning is wrong so long as you get the right answer! The worst thing of all is to follow the correct reasoning and end up with the wrong answer! This was a chastening experience, and it became my "little problem". Every time I met my manager, he asked how I was dealing with my little problem! I therefore kept telling him it was completely resolved. Eventually the matter was dropped.

In those days, computers were not as reliable as they are today. They frequently "went down". We used to have a set procedure to follow if the machine went down.

1. For the first 2 hours the local IBM site engineers (i.e., Tim Lloyd, myself and the IBM Customer Hardware Engineers worked on the problem.
2. For the next 2 hours local support (Liverpool) was called in
3. At 6 hours the Divisional Support was called in.
4. At 8 hours National Support was called in and the USA 360/65 Plant informed.
5. At 12 hours the Chairman of IBM (UK) was informed.

This procedure prevents Engineers from getting bogged down in the problem. Very often when the next level of support arrived, and you explained the problem to them you suddenly realised what was wrong! It was a good system. The last thing you wanted was the Chairman of IBM(UK) being got out of bed at 4am in the morning!

Complex Operating systems support multiprogramming, that is, more than one program running at the same time. Strictly speaking, in a multiprogramming system programs do not run concurrently. At any one time only one instruction is actually being executed in the computer and the operating system switches between programs. So, when one user program has to wait for some input or is sending data to a printer, the operating system will cause another user program to take over. In fact, there is only ever one instruction running at any one time and this is shown on the machine console in a set of 32 lights called the Program Status Word. This shows the current machine code instruction actually being executed. Of course, there will be many thousands of instructions operating one after the other and the Program Status Word (PSW) will just be a blur of lights. However, the machine can be stopped and PSW will show the current operation being processed at that moment.

On the right is a system 360/65 console. Across the middle are the lights where the PSW is displayed. This shows the current instruction being executed. You can also see the switches which enable a Systems Engineer to alter the Program Status Word.

Every so often, the Operating System is upgraded, and the process is called a SYSGEN (System Generation). Extra facilities are provided, and previous errors are corrected. Tim suggested that I should carry out the Release 16 -> Release 17 SYSGEN for MVT on my own to make sure I really understood the system.

SYSGEN is a huge operation, starting with a couple of week's preparations, and ending with the process of updating the system in front of the customer which takes about 3 hours. I nervously carried out the preparations and was finally ready for the update at 5 pm on Friday, 20 December 1968. I pressed the button with Tim Lloyd and all customer personnel watching. The SYSGEN commenced and immediately the phone rang. It was my wife, Mary. "I think I am going into labour. I am about to give birth. You need to come home straight away." I replied "I have another baby to deliver first. I will be home in about 2 ½ hours!" Although it sounded as if I didn't care, I reasoned that for our previous babies (Gareth and Carys), Mary had been in labour for 28 hours and 18 hours respectively, so she wouldn't be giving birth in the next three hours (thank goodness I was right!).

I completed the SYSGEN successfully. Tim commented that it had gone brilliantly, and I arrived home about 8.30 pm. Our second daughter, Cathryn, was born about 2.30 am and I was present at the birth. I was of course delighted that it was all over, and that baby and Mary were well. Many men had told me that being present at the birth was a wonderful experience, but I didn't feel that. I was just relieved it was all over.

In December 1968, Tim Lloyd shocked me by announcing that he was going to move to the IBM Education Centre in London the following summer. I wondered who would take over the account. I assumed that another experienced Systems Engineer would be drafted in.

Tim finally announced that he was leaving for the Education Centre in the summer, and I awaited who would be announced to replace him as site Systems

Engineer at Daresbury. To my consternation, I was called in by my Liverpool Manager and he said, "Congratulations. You are now the Site Systems Engineer for Daresbury!"

I replied, "I am not sure that I am really ready for such a key job."

"You will grow into it, and Brian Scarisbrook will be there to assist you," he replied. Brian was an IBM Process Control Engineer and didn't know a lot about Systems 360, but he was an experienced Systems Engineer. This again was typical of IBM. They would really push staff into higher level jobs and take the risk.

I had obviously impressed the people at IBM because, on 2 June 1969, I was promoted to a Full Systems Engineer at a salary £2448 per annum. I hadn't even finished my training schedule. It normally took 18 months in IBM to move from a Trainee to a Full Systems Engineer. I had done it in 15 months. The reason was not that I was exceptional, it was the quality of all those who had helped me like Tim, during the period.

My main location was still Daresbury where I was now in charge, but I began to be used increasingly to support MANWEB (Merseyside and North Wales Electricity Board) in Chester. They were interested in analysing their electrical network on the computer and IBM had a program called ECAP on which an engineer could draw circuits on a Visual Display Unit (VDU) and analyse them. Remember, this was very early days for VDUs, but IBM had developed the 2250 Graphics VDU, on which a light pen could be used, and Daresbury had bought one. I found it very interesting working with the Engineers on the 2250, which in those days was a very expensive piece of kit.

If a computer system crashes, the Systems Engineer needs to find out why. The Program Status Word on the main console shows the instruction where the crash occurred. The Systems Engineer can then obtain a printout of the complete content of the store. This is a huge print out since it is showing the instantaneous state of every bit in the machine (and there were 8 million of them in the Daresbury 360/65!). This huge paper output is called a "systems dump".

As already pointed out earlier, a Bit is a "0" or "1".and a collection of 8 bits is called a "Byte", A byte can contain a character (like "A", "B" or "3") so the word "hand" could be contained in 4 bytes. The basic unit in a computer is a "word", and the length of a word will vary with different computers but for the IBM 360 a word was 4 bytes (or 32 bits).

A huge set of 0's and 1's is difficult to read so in the print-out, each byte is divided into two four-bit parts, each of which can hold a number from zero to fifteen. The bit patterns 0000 to 1001 are represented by the normal decimal numbers 0 to 9, but how can the bit patterns 1010 to 1111 be represented? They actually represent the decimal numbers 10 to 15, but that would not work since we need a single new character for each, so they are represented by the letters A to F. The byte 11111001 would have a representation of F9, the byte 01001010 by 4A and a complete word (one instruction) i.e., 11110011101001110001101111110010 would be F3A71AF2. These are much easier to read than the individual bits. The complete contents of the computer storage at failure (each word of 32 bits) are printed on a paper listing using this notation—with 8 per line.

Opposite is a page of the printout of the store. It looks horrible, but once you understand the control blocks and what they mean, you can find out what is happening. Much of this information is of no interest (simply the content of the various user programs running in the machine) but at the beginning of the listings are masses of control blocks which tell the Systems Engineer what is running where, which task is active, which tasks are waiting etc.

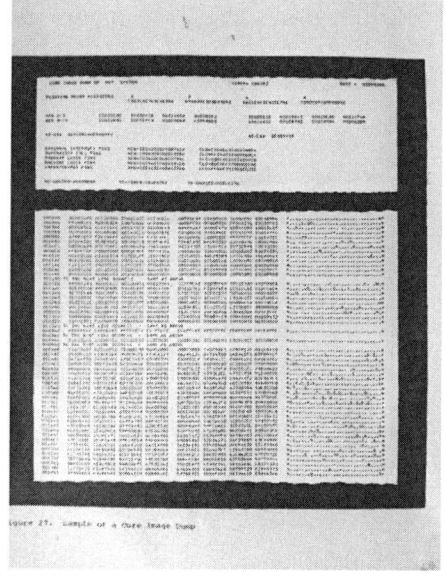

The control blocks therefore assist the Systems Engineer in finding out what was happening at the time of the crash. I developed the technique of sifting through this information following all the control blocks to find out what was wrong. It was complex work. One had to understand the basic layout and connections of all control blocks of the Operating System and follow the control groups and addresses. There is a Master Control Block in a fixed location, and this points to other control blocks to tell the Engineer what was running at the time, what the Operating System was currently doing, and what Input Output processing was happening. It's a bit like a treasure hunt where people in cars follow a set of clues to a prize. Each clue points you to the next clue. Although it looks formidable, with extensive practice you can eventually become quite good at it.

This skill proved its use in the spring of 1970 when the 360/65 started to do odd things. Initially there was an occasional break down, and I analysed the dump and informed the hardware engineers that there was a hardware fault at a particular location. They found there an open circuit and repaired it. A few days later it happened again, and again it was a circuit fault. The machine then began to crash more and more frequently and each time it required about an hour of my time to find the fault. When it began to happen more than once every day, we contacted the plant at Poughkeepsie (near New York) to ask them what on earth was wrong with our 360/65. They eventually came back with the answer—We know what's wrong and you won't like it! Your machine was slowly going open circuit!

It turned out that at the plant in Poughkeepsie where the Model 360/65 was manufactured, a worker on the production line had suggested that IBM use a different alloy of silver to solder the connections. This alloy was considerably cheaper, and the workman received a cash award for the suggestion. From then on, all subsequent 360/65 machines had been manufactured using this alloy, including our 360/65 at Daresbury. Unfortunately, the alloy turned out to be metastable. Over time this meant the constituents separated out and broke the connection in the machine. What we were seeing in the gradually increasing number of breaks were the disconnections on a diffusion curve as the alloy separated.

"How do I tell the customer that his computer is slowly going open circuit!" I said to Poughkeepsie.

They replied, "We are sending in three teams of IBM Hardware Engineers from Sindelfingen in Germany. They will work round the clock and rewire your machine."

"It will never work," I said with resignation. (8 million solderings!) I then plucked up courage and went to see the Director of Daresbury (Prof Zacharov).

He exploded. "It will never work again," he said, and he also couldn't believe it.

"Trust IBM," I said (but feeling distinctly worried).

The engineers duly arrived and rewired the machine in about two weeks. I had to re SYSGEN the system and the customer management gathered round the machine for the momentous event. I pressed the button and the machine started. After 2 1\2 hours it was obviously that the SYSGEN was going to be OK, and the Director virtually went on his knees and said, "Thank God for IBM!" He then turned to me and said, "IBM is expensive, but you get what you pay for. No other manufacturer could have done that." Did the man in Poughkeepsie keep his money—we'll never know!

On 13-17 April 1970, I went to the IBM Systems Engineering Annual Symposium in Amsterdam. One Systems Engineer from each region was selected to go because of performance in the previous year. It was a superb Symposium and was made even more interesting because the Apollo 13 Moon Landing Mission suffered a serious failure on 13th April. The mission had to be aborted and there was a serious risk of the Astronauts not being able to get back. The incident was made famous in the film "Apollo 13" starring Tom Hanks. For three days it was touch and go as to whether they would get back to Earth. IBM flew over an expert from NASA who told us the whole story as it unfolded. It made the Symposium a memorable event.

Early in September there was another problem with the 360/65 at Daresbury. The machine suddenly failed and using my dump expertise, I traced the problem to a particular instruction which transferred control, in error, to a wrong part of the store. The address where control should have been transferred was always fixed (It should never change) but it had been inexplicably changed. The system failed because the new address was a wrong address which then brought the system down. This sort of problem is difficult to solve because the change to the address could happen a few million instructions earlier and it cannot be determined where it happened. It is not usually possible to work backwards from the event to the cause.

It's a bit like being on a car treasure hunt, when you have to drive between locations and one location and find a clue to the next location, it is as if someone has put an incorrect location on the route. Once you reach the wrong location you are lost and cannot get back onto the correct route.

Luckily, the previous instruction to the faulty address instruction was known (It was the previous instruction). So, by using the switches I was able to display this previous instruction and I altered it to jump to a section of the store which was not being used. I then inserted an instruction which checked if the address in the faulty instruction was OK. If it was, I added an instruction to jump back to it. If it was wrong, I added an instruction which put the correct address back in and then added 1 to a counter so that I would know how many times the correction had occurred.

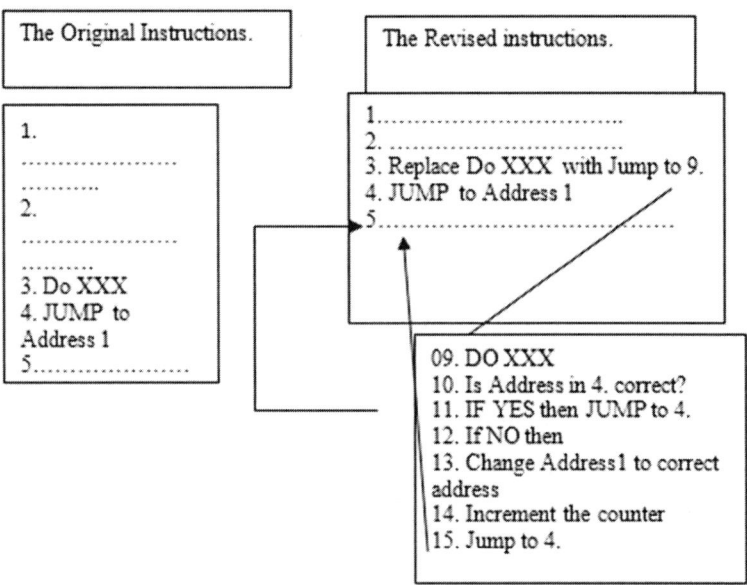

The sequence is shown above.

This meant that if the instruction was OK, nothing changed. If it was in error, it was changed back to the correct instruction and the counter recorded the number of times this had been done. Using the car game analogy, if some malicious person was changing the instructions at one of the locations, when people arrived at the instruction sending them to it, they were directed to an alternative location before the faulty one, bypassed it, and then carried on as normal.

This extra code meant that the machine kept running even if the address was changed and that I would know how often it might have failed. The next day I recorded about a 1000 corrected errors, but the machine kept running. We did not know why the address was being changed but after a few days that the number of errors slowly began to decrease and eventually it didn't happen again. We never found out what had been wrong—that's computers for you!

Another job of a Systems Engineer is to support the Salespeople in selling or upgrading machines. The IBM salesman for Daresbury, Chris Henfrey, with whom I worked, was keen to sell an upgrade to Daresbury in the Autumn of 1970. He wanted them to upgrade to a 360/85. He therefore asked me to do what is called a "Benchmark" to show Daresbury what the 360/85 would do for them. In a Benchmark the Systems Engineer takes a representative set of jobs from the customer and runs them on the proposed new machine. Since there was no 360/85 in the UK, we had to go to the IBM Systems Centre at Poughkeepsie (upstate New York) to carry out the benchmark.

I went with Systems Engineer (Brian Scarisbrook) who was really a process control expert so the whole thing was my responsibility. Benchmarks are quite complicated, and they must be set up with all the appropriate Job Control specifications for the computer. Also, since many other groups will be there doing benchmarks for other customers at the same time, there is considerable pressure on machine time. After three days of work at Poughkeepsie we were just about ready to run the benchmark overnight, when we had an urgent telephone call from IBM (UK).

It said that Robert Hall (another Senior Systems Engineer) was arriving that night from the UK and was going to do a final Benchmark for the UK Meteorological Office who were considering buying the 360/195 (right). This was a hugely prestigious order, and all depended on the results of the benchmark. IBM told us to drop everything we were doing and provide 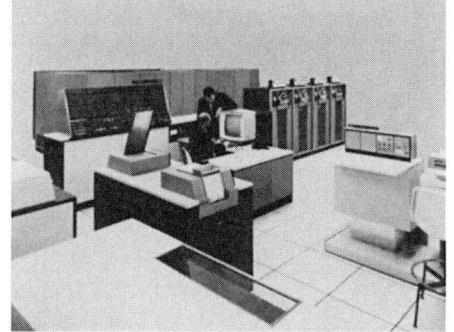 all necessary assistance to Robert. We worked with Robert for a week and finally sent the results back to the UK. Two days later Robert received a telegram. "The Met Office have placed the order—stop—Congratulations on a job well done—stop—Proceed directly to Bermuda—stop—wife and family await."

"What about us?" we asked.

"Carry on as normal," said IBM. So, Robert left for Bermuda, and since we had run out of benchmark time, we tidied up and flew home.

A few months later, we did the benchmark on a 360/85 which by now had been installed at the Systems Centre at Hursley in the UK so there was no need to go to the USA. We went down to Southampton and booked in at an hotel. The first couple of nights we had the machine overnight to set up the Job Control. At dawn we returned to the hotel and slept until the afternoon. Then at midnight we set up our benchmark on the 360/85 and started the run (which takes several hours). I had just poured myself a coffee at 3am and the Benchmark was proceeding nicely, when the door burst open, and a Senior IBM manager came in with two bleary-eyed Systems Programmers. "Get off the machine immediately" He said, "The Alitalia Reservation system in Rome has crashed and we have to solve the problem."

We did as we were told. A taxi was waiting for the programmers outside and a chartered plane at Heathrow for when they had solved the problem to take them to Rome. As I watched them pouring over the listings I thought "This is REAL pressure!" After about 2 hours they had clearly managed to sort the problem out and off they went to Rome. By now it was too late to restart the Benchmark, so we packed up and returned to the hotel. It was 9.30am and the Hotel manager curtly said, "you are too late for breakfast!" I replied that we had been working all night.

He suddenly remembered that I was "Dr Alty" and mistakenly thought we had been saving lives all night! "Don't worry Doctor, we will get you and your team a good breakfast." We thanked him but decided not to reveal our real identities and it was a very good breakfast! We never completed the 360/85 Benchmark and shortly afterwards, a new computer family, the System 370 series, was announced so Chris Henfrey decided to wait and offer a large 370 when it became available.

Computers do occasionally behave oddly! In March 1971, Harwell (who also had a 360/65) decided to change to the MVT operating system. I thought it was a great idea as it was a bit lonely running the only MVT account in Europe at Daresbury. They successfully installed the system but after a while the system started playing up and crashing frequently. I was called in as the MVT UK expert (!) to sort the problem.

The Senior Systems Engineer at Harwell was Keith Harraway for whom I had great respect. We worked together for nearly a week trying to sort the problem, but even after examining many system-dumps I couldn't locate a real problem. We tried many ideas without success. Then, on the Monday morning, we switched on the system, and it ran like a dream. On Tuesday and Wednesday, it kept running, and the customer was delighted. I pointed out quietly to the Salesman that we hadn't actually found a fault. "Don't tell that to the customer" he replied, "You have fixed it and that's all that matters."

The Head of Computing (Dr Saddler, I think) called me in and expressed his gratitude. I bit my lip and smiled, and the Salesman gave a look which said, "Don't you dare open your mouth!" We never found out what had been wrong. In our efforts we had obviously fixed the problem by accident, but the machine ran fine from then on—computers are sometimes like that! However, the incident did bring back to my mind the comment of my manager in 1968 —"What matters in IBM is getting it right—we don't care if your reasoning is wrong so long as you get the right answer"!

I really enjoyed working for IBM but in April 1971, I saw an advertisement in the Sunday Times for the post of Director of the Computing Laboratory at Liverpool University. It was a Professorial level position and I suspected that I was not qualified enough (as I was only 31). However, I contacted Prof James Cassels (the Head of Department at Liverpool Physics Dept.) and after some thought he said, "Why not apply, in the worst case you can only lose." I did apply and was shortlisted. The interview took place on 9 June and on 11 June, I received a letter stating that "The Selection Committee has decided not to make an appointment to this post for the time being and I regret, therefore, that your application must be considered unsuccessful." I was not over surprised and didn't think much about it. I was still rather young. However, this Liverpool application had an unexpected effect one year later as you will see.

Late in 1971, I again went to the Annual Systems Engineering Symposium in Madrid. Shortly after I returned an event occurred which would have a major effect on my work at IBM (although at the time it didn't seem relevant). On 30 June 1971, IBM announced the successor range of computers to the 360 range and they called it System 370. IBM, as usual, claimed that they had prized an extra 10 degrees out of the compass! Everyone in the office had gone down to London for the 370 announcements. Since someone had to be in the office to handle customer calls, I volunteered to stay in the office and man the telephones.

The telephone rang, and I took the call. It was the Treasurer of Flintshire County Council (who was a current IBM customer with a small 360/20 installed). He said "I would like to order a.... (and then there was a rustle of paper) ... 370/145 for Flintshire County Council." I thanked him for the order and assured him I would pass it to the salesman (a person called Tony) which I did. Tony couldn't believe it. He said "The upgrade path for a 360/20 computer is to a 370/135, and even that is rather a big leap. He must have got his numbers wrong!"

That week, Tony went to Flintshire and saw the Treasurer. He pointed out that the 370/145 was too big for Flintshire requirements and that the 370/135 would be the appropriate upgrade (and cost about half as much!). The Treasurer looked at a piece of paper in his hand. "No, I want a 370/145" he said. Tony couldn't understand this, so he went to the Computer Manager at Flintshire who agreed that the request was wrong. Both worked on a 370/135 bid, costed it out, and then eventually went back to the Treasurer, in September, showing him a saving of about £200,000. The Treasurer didn't say anything, but when they had gone, he picked up the telephone and called the Chairman of IBM (UK). "When I say I want a 370/145, I mean it" he said, "Get that salesman off my patch, I do not want to see him again!" Tony was immediately moved to another part of IBM.

At this time our Manager at the Northwest Marketing Unit was changed and a new manager, Ernie Cochran, took over. Ernie called me in. "You know what's happened at Flintshire. We would like you to take over as Salesman!" Since I had just been promoted to Senior Systems Engineer, I was again gob-smacked!

On several occasions IBM had suggested that I might change to sales, but I had resisted. Ernie said to me "You are a good Senior Systems Engineer, but they are two a penny(!). Good Salesmen are rare breeds and worth their weight in gold. We think you will make an excellent Salesman."

"But I have not even been on Sales School," I replied. Sales School was a tough final passing out school for all Salespersons in IBM.

"That can happen later," he said "Firstly sort out the mess at Flintshire. Oh, by the way, you are also Salesman for Chester and Denbighshire as well."

I accepted the post and said, "If Flintshire want a 370/145, that's what they will get." Of course, being a salesperson changed the way my salary was paid. Salespersons were paid partly on commission. You could either have 60/40 (i.e., 60% basic and 40% commission) or 80/20. Cautiously, I opted for 80/20.

I went back home, puzzling why Flintshire wanted such a large computer—there must be a reason. I talked with Mary my wife. Could there be a reason connected with Wales? Suddenly it hit us! Under local Government Reorganisation proposals, Flintshire and Denbighshire were to be joined into one county—the county of Clwyd. Clearly Flintshire were stealing a march on this. By getting a 370/145 ahead of the amalgamation they would take over computing for both counties! I immediately placed an order for the 370/145 and it was due for delivery in mid-December 1971.

I knew that eventually Denbighshire would find out and would be upset (and, of course, particularly with me as their Salesman). It happened on a train coming back to Liverpool from London, probably in October. I was sat with the Treasurer of Denbighshire in the Dining Car having dinner. The first-class carriages had two seats one side and four seats the other side. Opposite us were three programmers from Flintshire. Suddenly one of them leaned over and said triumphantly "We are getting a 370/145!" The Treasurer of Denbighshire was thunderstruck. "Is this true" he said. "Yes" I replied since I couldn't think of anything else to say! "This is terrible" he replied "How could you do this to us. See me in my office tomorrow morning."

I met the Denbighshire Treasurer the next morning I pointed out that each account was confidential and that I could not divulge to him what another County Council was planning. "What would you have wanted me to do if you, not Flintshire, had thought of this?" I reasoned. "You would not have wanted me to tell them." I pointed out that I had to respect either side and keep their actions confidential even though I might see one taking advantage of the other. Eventually he calmed down and realised that he was the person who had missed a trick and he couldn't blame me for it. I said to him why not upgrade to a 370/135 and he brightened a little!

I rather enjoyed Sales but there can be serious pitfalls in the job. Whilst planning proceeded for the installation of the 370/145, in November, Cheshire County Council (another of my customers) cancelled a modest Storage upgrade to their system but ordered 12 x 2319 Disk Drives. These were new high-capacity disk drives recently announced and were being ordered by everyone. I put in the change order to IBM for the new disks and cancelled the storage upgrade.

On December 14th I began to feel ill, but I still went into the office. Ernie called me in and pointed out that as a salesman I must carefully check the NIRI-GIRI List for my customers. I didn't know what this list was, and I still don't know what the NIRI-GIRI acronym means (and I have probably spelled it wrongly!). Essentially it is a list of all the current installed and on-order equipment for a Salesman's customers. Ernie said it was important and he would insist on me reporting back that I had done it properly. Because I was feeling ill I did not check it, and this was an expensive mistake which became clear in the middle of January.

On 17 December 1971, the IBM 370/145 which I had sold to Flintshire was installed. It was a great occasion to which I, naturally, was invited. However, I began to feel ill the night before the installation. I felt awful, and when I awoke on the Monday morning, I couldn't get out of bed I was so ill. I had Flu! The doctor came and told me to stay in bed for about a week. I said to him that I had heard that Whisky was good for Flu, and would he recommend it? "It won't do you any good, but you might enjoy it" he replied. I was very ill over Christmas.

Flu is much more serious than a cold. By this time Mary had become ill as well so I don't think we had a Christmas Dinner that year and I remember a neighbour leaving soup on the doorstep!

Flu is much worse than a cold and it affects you for some time. Between Christmas and New Year, I still didn't feel very well but I went to see the DP Manager of Cheshire County Council. He told me he was thinking of ordering five Visual Display Units from a competitor. This is bad news for a salesman because, once a competitor gets an order into an account, he can call continuously trying to sell other equipment. I was not well that day and the terminals were half the cost of IBM equivalents. "Is there any reason why I shouldn't order them" he asked. In my muddled and ill state, I couldn't think of any.

When I returned to the office, Ernie Cochran, my manager, went spare. "You have let them in" he said. "You should have told him to wait whilst you checked what IBM could offer. You had the whole expertise of IBM to help you and you didn't use it." I went back to Cheshire, but they had ordered the terminals.

Things went from bad to worse. The manager from Cheshire County Council who had ordered the 12 x 2319 high-capacity disks for delivery in April 1972 (and at the same time had cancelled an order for more store) asked if the disks could be installed at a weekend to minimise disruption. I called in at the Liverpool Engineering Support Office to see if this could be done. Liverpool checked the installation schedule and pointed out that there were no disks on order! I couldn't believe this and checked with Head Office, and it was true. A person in the IBM order department had misunderstood my order change in November and had cancelled both the store AND the 12 disks! This was serious because all IBM equipment is ordered strictly in a sequence. No customer can take priority and these disks were like gold dust! Everyone wanted them.

Remember, I had been asked to check the NIRI-GIRI list in December by my manager. If I had done this, I would have found out that the disks had been wrongly cancelled. This was a serious error. The Disks were extremely popular and there was, by now a serious backlog of orders for the disks. Cheshire wanted the disks because of the Local Government re-organisation that was taking place in April. I went to Ernie and confessed what had happened. He initially said, "Right, you are really in trouble now and I will discipline you for this. You have a problem!" However, I did point out that he had insisted that I should report back to him when I had checked the list and he hadn't done this—so he was in trouble as well! His, "You've got a problem" changed into "We've got a problem"!

The customer went spare when I told him. Over the next few weeks Ernie and I worked on the problem, trying to hunt down cancelled disks. Eventually we managed to install six 2319 disks on schedule and six, three months later, but we had to compensate Cheshire for the late delivery of the second set.

Early in February, I was still smarting over the Cheshire order for six competitive terminals, but there was no way I could propose anything cheaper. Indeed, the VDUs that IBM offered were a lot more expensive and, frankly did not offer anything more. What happened next, however, was an interesting exercise in salesmanship. I kept asking why Cheshire wanted the terminals and I was told it was the Highways department that needed them. I called at the Highways department and asked what they were using them for, and they responded by saying that the main advantage of the terminals was a tape cassette recording facility. I couldn't see why that was useful and kept asking why it was important. Suddenly the customer replied, "Because of the Soil Laboratory." I immediately replied, "Take me to your Soil Laboratory!" I had not realised that such a Laboratory existed.

The Soil Laboratory was a large room with a huge data collection facility in the middle (very old fashioned) which produced data on cassette tapes. They wanted the terminals so that they could feed in the cassette tapes to the main computer from these terminals, and then store results from the main computer back on the cassettes. I pointed out that this was really antiquated and what they really needed was an IBM System 7 Process Control computer in the Soil Laboratory, which was connected, directly on-line, to the main computer. Since this cost about £50,000 Cheshire thought this was an outrageous idea and even people in IBM didn't believe that it was a solution that Cheshire would accept. However, I was right. Six months later, they ordered a System 7! This was a classic example of good salesmanship—instead of criticising them for ordering the competitive terminals, you must listen, and by listening I discovered the Soil Laboratory, and eventually offered a much better solution.

I mentioned earlier that I had never been on Sales School, and IBM kept suggesting that I should go. Sales School is an intensive course of sales training lasting two weeks and it is the final passing out course for salespersons. Every time they organised a two-week booking for me on Sales School, I managed to get customers to organise an important meeting and said I couldn't go! However, in February, IBM outwitted me. They organised my attendance on Sales School and brought in a salesman to cover for me for the two weeks.

I went at IBM Sales School in the last two weeks of February 1972. It was an amazing experience and really intense, I learned a lot which helped me in later life. They taught some really interesting sales techniques. There are many salesmen (time-share etc.) who con people into buying things they don't need but I realised that high level selling was much more ethical. Customers really did need to be guided into buying the right equipment. They often had the wrong idea about what a device could do, or how it might benefit them. I remember being taught the key principles.

1. Describe the features of the new equipment (i.e., tell them what you are offering)
2. Importantly, convert the discussed features to benefits. (Features are only worthwhile having if they provide a benefit to the customer)
3. Assume the sale has been made and ask for the order. (This is not actually expecting the order to be placed, but it forces the customer to react if they have objections).

The second point is very important. Many customers talked about features, but it is benefits that make the difference. Cheap Disk space might be very cheap, but do you need it? The last point is crucial and is designed to make the customer think clearly about the proposal. Many customers will agree the benefits, but when you ask for a decision, they become hesitant and keep thinking of problems. There is nothing wrong in that, but they must think it through.

When the salesman asks for the order, he does not normally expect to get the order. The statement jolts the customer into focussing on what the problems are. So, when the customer, as is usual, reacts strongly to the order suggestion, the salesman must follow this with a question which cannot be answered by "yes" or "no" (a question beginning with why, what, when, how, who!). This then digs deeper into possible customer objections. Eventually the salesman will get to the real objection. Then it is either an objection which is unfounded, or a real one which needs to be overcome by other benefits. The System 7 sale to Cheshire County Council I mentioned earlier, was a classic example of the technique being used successfully and to the benefit of the customer.

At the end of the Sales School the attendees were graded and the person who came top had to propose the health of IBM at the final dinner in the presence of the Marketing Director of IBM (UK). Those who didn't pass the course were not invited to the dinner! I actually came top of the Sales School.

On the final Thursday night, the 23rd February, I gave the toast and Peter Morgan (the Marketing Director) replied. They said my career was now made in IBM. That night we had a great party afterwards. Although I was delighted to come top, I did have an advantage over many of the other attendees since I had worked in IBM for 3 years, whereas they had only been in IBM for 18 months, so it wasn't quite as outstanding as it sounded.

The next day I was exhausted and travelled back home to Knutsford. I was so tired I went to bed at about 7pm and I remember the telephone ringing at about 8pm. It was Prof Cassels from Liverpool University. He said "You may remember that about a year ago we interviewed you for the job of Director of the Computer Laboratory at the University and we turned you down because we thought you were too young. Well, we realise that you are now a year older, and I have been authorised by the Vice Chancellor to offer you the job! It is a professorial level position. Come over on Sunday and have lunch with me and talk it over."

This was an amazing offer, and it was odd that it came one day after I came top of Sales School which was regarded in IBM as a step which could make your career. I spent Saturday thinking about it and decided it was too good an offer to turn down. I saw Prof Cassels for lunch and indicated I would accept. On the Monday I went into the office and was greeted by Ernie. "Congratulations. A wonderful achievement for the Unit" he said and pulled out a bottle of champagne. "Before you give it to me, I must tell you that I am leaving" I replied. Ernie quickly pulled back the bottle of champagne!

IBM couldn't cope with this. I was sent to see the regional director (David Livermore who was great guy and a classics scholar). IBM had obviously done their homework because he said "you do realise that Liverpool have been turned down by the Computer Board for a 1906A computer and are now getting a 1904S? The Director who died last year was a very poor Director." I replied that didn't know this, but I said "Liverpool is a very large University. It obviously should have a 1906A and I think it is possible to reverse the decision." Eventually I had to fly to London to see the Country Sales Director, Peter Morgan. He offered to send me to do research at the IBM Research Centre in San Jose, but he was very nice about it and realised I had made up my mind. He wished me well and I accepted the Liverpool offer to commence as Director of the Computer Laboratory on 1 April 1972. I left IBM for Liverpool University three weeks later.

A few years later, IBM invited me to a 2-day conference in the De Vere Hotel in Coventry. David Livermore was there (he had been the Regional Sales Director when I left). In the evening there was a Conference Dinner, and I was assigned to his table. After dinner, he gave a welcoming speech, and I initially was not really listening. Suddenly I heard the phrase, "he was a salesman with IBM" I listened intently and realised he was talking about me. He continued "Before he left IBM, he had come top of Sales School and on return to his branch his manager welcomed him back with a bottle of Champagne. When Jim told his manager, he was leaving, the Manager, quite rightly, quickly withdrew the Champagne" He looked at me, paused, and said "Jim—Here is your bottle of Champagne!" and he handed over a fine bottle of bubbly!

Chapter 11
Computing Director, Liverpool; I Discover HCI

In March 1972, I was appointed Director of the Computer Centre at Liverpool University. This Computer Centre served the whole University. The appointment was a Professorial Appointment, though my title was Director. This was a move from being involved in Digital Processing at the sharp end (i.e., IBM) into management of Digital Facilities, staff and inevitably into university politics! Also, my work would become much more teaching and research orientated.

The main functions of a university are Teaching and Research but to do this they require support services such as a Library and a Computer Service. Libraries were established from the outset, but the Computing Services only began to be established in the early 60's. Often they were originally attached to a department (such as Mathematics) but eventually became departments in their own right. Some were originally called Computer Laboratories though Computer Centre is now the normal term. With the publication of the Robbins Report of 1963 Universities grew considerably in size and new Universities were created so the computing load grew dramatically. At this time, the computers installed in Universities were decided by a Governmental Committee called the "Computer Board". Universities applied to the Board for Computer Facilities and the Board tried to solve this provision issue as best it could in a fast-moving environment.

Liverpool University had originally installed a DEUCE computer in the late 1950s (this machine was based upon Turing's original design) and I used it in 1963 during my PhD. In 1965, an English electric KDF9 replaced the DEUCE. It was originally housed in the Mathematics Building but later had a new building created especially for it.

In 1972, the KDF9 was getting old as Liverpool had been one of the six original "KDF9" Universities. Most of the others (Leeds, Nottingham, Sheffield etc.) had already replaced their KDF9s with ICL 1906A systems. Only Liverpool and Glasgow were left. However, Liverpool had a problem. The Computer Board, which allocated computers to universities, had recently instituted a new Regional Centre policy. They had installed large systems in Manchester, Edinburgh and London Universities and any remaining universities which required a new computer were required to get much of their large-scale computing from the Regional Centres and this would result in much smaller in-house systems. It was clear that Liverpool was the first real test of the Computer Board's new Regional Computing Policy and instead of a 1906A, they offered Liverpool an ICL 1904S to replace the KDF9. The 1904S had about half the power of a 1906A. I suspected it would be difficult to change their mind!

Before starting the job, I needed to get the general drift of what had been going on at Liverpool. Two weeks before I took up the appointment, therefore, I went to Liverpool to have lunch with the two current Assistant Directors of the Computer Centre, the Operations Manager, and the Education Manager, who had run the Laboratory for a year since the death of the previous Director. They told me how inadequate the previous Director had been and that he had not stood up to the Computer Board when they downgraded the bid from a 1906A to a 1904S. I asked what was happening about the Manchester Regional Centre usage and they told me it was very low, but that they were about to install an improved Remote Job Access computer the following week which might change things. Still, their main worry was that if usage on the Regional Centre was successful, this would prove the Computer Board's case for installing the smaller 1904S at Liverpool.

I pointed out that there must be, at Liverpool University, researchers who needed the power of the Manchester Computer for example, Chemistry, Engineering, Physics etc. and it was strange that they were not trying to use the system. I suspected that there was a considerable pent-up demand. I went home and thought hard about this apparently contradictory situation. On the one hand the Regional Centre must offer real power to some users – why were they not using it, but I also realised that if they used it this might justify the Computer Boards regional policy!

I started work as Executive Director of the Liverpool University Computer Centre on 4 April 1972. I had a staff of about 40, including 2 Assistant Directors, a Software Manager, Computer Officers, a Software team and Operators. We provided a Computer Service for the whole of the University. I first, therefore, went to talk with departments about their needs. The Physicists had given up using the KDF9, and long ago had installed their own IBM system, but the Chemists and Engineers were desperate for more computer power. It was very clear that there was a real demand for the Regional Centre power, but that the users were finding it difficult to submit jobs. Furthermore, the Computer Centre was not really being very helpful.

My main reporting line was to the University Computing Committee in the University. Several Professors were on the committee including Prof James Cassels who had originally contacted me about the job. I therefore went to lunch with Prof Cassels who told me about various political issues. He explained that the University were quite upset about the 1906A rejection, and they were looking to me to sort out the problem! There was also a proposal to extend the Computer Centre Building for the new computer, but this assumed that a 1906A would be installed.

Because of the importance of Liverpool's role in the new Regional Centre, I was invited to become a member of the University of Manchester Regional Computer Committee (UMRCC) and on 21 April, I attended my first meeting of that Committee in Manchester to find out how they operated. Each Regional Committee had a Board Member as a representative from the Computer Board to ensure fair play for the regional University users. I was surprised to see that the Board Representative was not there (and this had ramifications later!).

Whilst I thought that the Manchester system would be fine for really large jobs, I felt that the idea of downscaling the local computing of a university the size of Liverpool to a 1904S was ridiculous. Liverpool was a large University with many successful departments. The School of Tropical Medicine was famous world-wide. There were no less than 17 Fellows of the Royal Society in the University (many Universities only have one or two) and the Medical School, Chemistry and Physics had international reputations. One of the Professors of Physics (Sir James Chadwick) had won the Nobel Prize for Physics for discovering the Neutron, and Profs. Clarke and Shepherd, of the Medical School, had cracked the Rhesus Negative problem and saved the lives of many thousands of young babies.

I spent some time thinking how to sort the 1906A problem out, and after talking with the users I realised that there was even more demand for access to the Regional Centre than I had first thought. I therefore went down to meet with the Secretary of the Computer Board—Lionel Rutherford—to discuss the situation with him. I pointed out that Liverpool was a large University and, even with Regional Centre usage the 1904S would not be powerful enough to take on the local load. He agreed that might (in the long term) be true and that we could come back for an upgrade later. He pointed out that the Board expected about 50% of the total load to be sent to the Regional Centre.

Suddenly, it was blindingly obvious what to do—give the Board what they want! I had been taught in IBM that it was important to always give someone what they wanted but make sure you also got what you wanted as well! Therefore, I should show the Board that their Regional Policy worked but make the load so great that they would have to upgrade our bid to a 1906A as well. I decided, therefore, to really encourage use of the Manchester system.

I had already identified that the staff at the Centre were not really used to submitting jobs to the Manchester 7600 (see right) and that this was probably the main reason that our load was so poor. The users did not understand how to submit jobs and the Computer Centre wasn't really helping them. I decided, therefore, to really  encourage use of the Manchester system, to such an extent that I would "break its back!"

We had been allocated a 12% share of the Manchester Computer and my plan was to saturate it by August. The first task was to make sure all the staff of the Centre were competent at using Manchester so that they could pass on this expertise to our users. I therefore insisted that ALL programming and advisory staff of the Laboratory should submit jobs to UMRCC on a daily basis so that they would get used to it and then be able to help our users. I had a weekly report given to me on staff usage and I put pressure on all staff to do this. The idea was that once the staff understood how to use UMRCC, they would be able to advise users and usage would grow. The jobs submitted by staff were not large, so they did not increase the load a great deal, but the staff learned rapidly how to best use the Manchester machine.

The Board had already told me that they expected an offload of about 50%, so if the load at Manchester could be developed sufficiently this would mean an ICL 1906A at Liverpool as well. Of course, I did not know whether such a load was possible, but the users had told me that there was a great deal of pent-up demand because the KDF9 was running out of steam.

Submission started slowly at first but then it began to rise rapidly. We made no attempt to artificially load the system. I had little idea which users were beginning to use the system, so it was a totally natural load. However, my intuition about the pent-up load was correct. By the end of July, the usage had rocketed.

By August we had exceeded our allocation of 12% at UMRCC and UMRCC said they were going to restrict our access. I retorted "You ain't seen nothing yet" and I ignored them! I immediately reported this to the Computer Board, pointing out that our load at UMRCC was now equivalent to half an ICL 1906A (i.e., a full 1904S workload). The Board were in a fix! We had proved that their regional policy worked and so they were very grateful. They had suggested that our off-loaded work would be 50%, but this 50% was inadequate for a local 1904S.

In September, I made a brand-new case for a 1906A but before I submitted it an event happened which could have derailed it. All Computer Directors attended an annual conference called the IUCC (Inter University Computing Committee) every year. In September 1972, it was held at the University of Surrey, and I attended for the first time. I gave a paper which illustrated how our concentration on the user interface had led to the explosion in usage of UMRCC.

The paper was well received and The Chairman of the Computer Board, who was there (Prof David Finney), congratulated me on our regional usage. On the second afternoon, Prof Finney gave a talk on Computer Board Policy and stated that one key feature of the policy was that a Board Member sat, as of right, on the Regional Centre Management Committee to ensure that the rights of the regional users were properly served. The current member of the Board on the UMRCC Committee was very well known in computing circles, but by September I had attended at least five Regional Committee Meetings at UMRCC, and this representative had only been at one of them!

I thought this was a bit much, so I rose to my feet in the open session and pointed out to Prof Finney that it was an excellent idea but did not work if the Board Representative did not attend the Committee regularly, and this had happened at UMRCC. I did not mention the representative by name, but he was there at the meeting, and immediately rose to his feet outraged by my intervention saying I had insulted him! Total pandemonium ensued, and we then fortunately had a tea break. A senior ICL man came to me and "you may have blown your case for a 1906A" and I must admit I was a bit worried, but I sought out the Chairman, Prof Finney during the interval, and he agreed I had made a very good point. It was essential that Board representative attended most of the Regional Centre Meetings. He would have a chat with the representative. The intervention probably actually enhanced the case for the 1906A!

Much later, I got to know the representative well. His name was Jack Howlett, and he was a really nice person and probably had good reasons for not attending, so in later years felt a bit guilty about raising the issue in public. However, if a Computer Board pushes a policy of full representation to support their regional policy (a policy which can seriously affect what computers were installed in a university) and they make public statements about it, then they are entitled to be challenged if the implementation is not working properly.

The Computer Board considered our revised bid in November. We had made progress because they suggested that the 1904S would be an interim solution, with us moving later to a 1906A. We went back and said this was not a useful solution, since it would cause additional expense and that we were already processing the equivalent of a 1904S at Manchester. In spite of still not knowing the outcome of our bid, in December I assumed we would win and commenced the plans for extending the computer building. In January 1973 the Board finally approved our 1906A, for installation in February 1975. It was my first major victory as Director and the Computer Committee were impressed. We opened the tenders for the new building on January 22nd and planned to begin the building work for the extension in June (The 1906A was a much larger computer than the KDF9).

Just before Christmas 1972, the Professor of Genetics approached me and asked if I could provide some additional computer power over the Christmas vacation. My plan was to close the computer on December 24th and reopen on January 3rd. The Professor's name was Phillip Shepherd and I had not met him before. I did not feel I should ask the operators to give up some of their Christmas holiday, so I initially said no. He was very disappointed, and he explained why he had made this request. He pointed out that he had acute Leukaemia and that he had only weeks to live. He was anxious to complete his research before he died. I quickly made some enquiries and it turned out that he was a very famous Geneticist, who had worked with Prof Cyril Clarke in cracking the Rhesus Negative problem. As a result, many thousands of babies' lives had been saved through his work. I discussed this with the operators, and they agreed to come in and provide additional computing. He was very grateful. I later found out that in the January he had gone into remission for a short while, but he died in October about 8 months later. It was quite upsetting at the time. He was world famous.

As well as retrieving the 1906A for Liverpool, this work had another very important effect on me. I realised that designing computer interfaces so that users found them easy to use was very important. The users were not using the Manchester Machine *because they did not know how to use it!* Furthermore, the staff of the laboratory were not familiar with it either. By forcing staff to use the system it made them realise that it was not too difficult to use when you *understood the interface*. Although improving the knowledge of the staff so that they could better advise users was important, a much better long-term approach would be to improve the design of the interfaces themselves. Current program interfaces were clumsy and assumed that users understood how the computer worked.

This taught me that the design of the *User Interface* was crucial in systems design and this resulted in a life-long interest in *Human Computer Interaction (HCI)*. I suspect that winning the case to upgrade to the 1906A at Liverpool might have been the first ever successful use of HCI with a positive monetary outcome. Later on, HCI became really important in making computer applications acceptable to a wider audience and vitally important in safety critical situations such a controlling aircraft or large nuclear power stations. This is covered in succeeding chapters.

It is interesting that coming into the University from outside can also have unexpected benefits. Ignorance of the way in which Universities work can be quite useful! This happened in the Spring/Summer of 1973 when we were busy planning for the 1906A upgrade, and I therefore had regular management committee meetings with the senior staff of the Centre.

At one meeting, I announced that the University had asked us to submit our equipment needs for 1973-74. I had never done this before, so I produced what I thought was a reasonably modest list of equipment requirements costing about £270,000 and the department management committee approved it without any comment. I thought it was odd that no-one suggested any changes. Later, I received a reply from the Liverpool University grant awarding body. They had approved about £230,000. I was quite pleased and pointed out to the management committee that they should not be too disappointed because we must accept that there would have been other strong competitive bids.

The whole management committee went into fits of laughter. They pointed out that in the previous year their bid had been about £30,000 and they had said nothing at the previous meeting because my bid was so outrageous! It is an interesting example of the fact that not knowing previous history can act to your advantage. If I had known that I would probably have put in a more modest bid!

In June 1974, an event happened which was not related to computing but was quite an interesting one. The family went to a Shooting Lodge (out of season) in Glenure, between Oban and Ballachulish in Scotland. It was easy to reach (about 4 miles up a narrow but good road from inner Loch Crearan), and it was a lovely house and location. Furthermore, the pub (the Creagan Inn) was only 4 miles down the road by the loch. The first day we arrived, the person who looked after it showed us round. All rooms had been modernised except one. When she showed us this room (which still looked very Jacobean) she shuddered and said, "It's a terrible room!" and we did not know what to make of this.

The following night we all went to the Creagan Inn (about 4 miles away) and deposited ourselves in the tiny "snug" bar. The bar was in the middle between the snug and the local bar which we could see through the bar opening. The lady behind the bar served us without comment. On ordering the second round she asked me where we were staying. "At Glenure Lodge" I replied. She looked a bit odd and turned to the local bar saying, "They're staying at Glenure!" The locals made a curious non-committal noise. Then she turned back and asked, "How are you finding it?" I replied it was quite nice. She turned to the local bar and said, "They're finding it just fine." The locals again muttered ominously. Finally, she said "Have you seen anything?" I didn't understand this but replied "Nothing unusual." She turned to the locals and said surprisingly "They haven't seen anything!"

The locals then made another even more curious sound and she turned back to us exclaiming "It was a terrible murder. There are bloodstains on the floor!" We now understood why the cleaner had said that the room was so terrible, and we assumed that the murder had happened there. When we got back to the house we rushed into the room and there did appear to be faded bloodstains on the floor, hidden under a mat!

It turned out that the murder was well known and was called the Appin Murder. We decided to investigate the murder and we found a plaque on the road to Ballachulish commemorating the murder which pointed in to Lettermore Forest (where the murder was actually committed).

What happened was this: The Master of Glenure (Colin Campbell, known as the "Red Fox") was a Tax Collector and frequently rode from Glenure to Fort William through Letter more Forest. It is a morning's ride. On May 14th, 1752, he set off and rode to Fort William after crossing Loch Leven on the ferry. On his return the Ferryman warned him that people were waiting for him in the forest, but he carried on. He was shot in the forest but managed to get back on his horse and the horse brought him back to Glenure Lodge. That night he died on the floor of the old room (so he was not actually murdered there).

Earlier, the Campbells had confiscated land from the Stewarts, and it was therefore assumed that the murderers were Stewarts. James Stewart (James of the Glen) was arrested and tried at Ballachulish. The Judge was a Campbell and the Jury were Campbells. He was found guilty and hanged at Ballachulish on 10 November 1752. On the scaffold he protested his innocence and recited Psalm 35 (which is known in the highlands as "The Psalm of James of the Glenns"). His body was left to hang at Ballachulish for 18 months.

Further study revealed even more interesting information. This shooting of the Red Fox in Letter more forest was actually used by Robert Louis Stevenson as the basis for the novel *Kidnapped* and the novel has a scene in which David Balfour and Alan Breck meets the Red Fox in Letter more Forest and ask him directions, when a sniper shoots the Red Fox. David Balfour and Alan Breck Stewart (commemorative statue in Edinburgh right) are then suspects of the murder. So, there were no ghosts but just a very interesting history with a literary conclusion.

Early in 1975 things went even better at Liverpool than we could have hoped. The building extension was ready and the 1906A was about to be delivered when the Computer Board Secretary rang me. He said that Glasgow had rejected their proposed 1906S (I think they wanted IBM), and he asked if we would take it instead of the 1906A. I immediately said "yes". It was a machine at least 50% more powerful than the 1906A, and so further vindicated my stand against the 1904S. So instead of a 6A we installed a 6S. This was an interesting machine with full virtual paging and running the GEORGE IV Operating system.

A few days before the end of the Board 1975 financial year (I think April), I had another telephone call from the Board saying they had a further £100,000 which they could not spend by the year end. If they did not spend it, it would be taken from them. I immediately rang ICL and ordered an extra high-speed printer and some additional disk drives and eagerly accepted the money! The Liverpool computer was now the largest and most powerful of the 1906 series in the world, so I really had changed the Liverpool computing situation in my three years as Director. The Computer Board now asked us now to provide remote job entry facilities for the Universities of Keele, Lancaster, and Salford, so we had even become a sort of mini-regional centre.

In March 1975, the 1906S system went in smoothly at Liverpool. After a few weeks of trials, we began running the system. We had also installed the Remote Job Entry (RJE) systems at Lancaster, Keele and Salford Universities. These were all connected via an ICL Networking System called GANNET across the four Universities to facilitate high speed networking. Across the Liverpool University campus, we installed RJE systems in Physics, Electrical Engineering, Arts and in the Administrative Building. In the RJE stations we used the University of Waterloo "Cafeteria" systems to give students quick access to the system for program development. Users would put in their card deck at one end of the Entry Station and slowly walk round to the printer, by the time they got there the output was waiting for them. It worked really well.

I pioneered the development of the networking system to give better connections between the four North West Universities. It was a system developed by ICL. The idea was to share resources and minimise cost. At first, whilst the network worked fine, the sharing didn't. All sites seemed to want to be the provider of services not the receiver. I was Chair of the Network Committee and was rather disappointed that sites would not share. However, after about a year, things changed. Suddenly sites realised that by using other sites for facilities it saved them a lot of work and the opposite situation occurred. Everyone wanted to hive off work to other sites! In the end a happy medium ensued with sites providing some services and sharing others.

At this time, I was asked to be an expert witness in a computer case. A large steel firm in the midlands was suing a computer supplier because the machine had not lived up to expectations. I submitted a report which concluded that I thought the Steel firm were correct. The action began at the High Court and was a very expensive case—3 QCs on either side. For the first three days we all went through the Ladybird Book of Computing (for Children) so that the Judge would understand! On the fourth day, the QC for the steel firm told the Judge that I had pointed to the inadequacy of the Systems Documentation in my report. He commented that he would show that my conclusions were incorrect. The Judge asked the QC what he meant by Systems Documentation. He tried three times to explain it and then said, "Your Honour, I haven't the faintest idea", to which the Judge said, "Perhaps Dr Alty can tell us!" I did so and shortly afterwards the case suddenly stopped, and the two parties came to an agreement. The Judge had been the Prosecutor in the Great Train Robbery Trial.

My work at Liverpool had obviously impressed the Computer Board because, in September 1976 I was invited to become a full member under the Chairmanship of Sir Henry Chilver. All major decisions on computers for Universities and Research Councils were taken by the Board. It had 8 members, 2 secretaries and several observers from various government departments. It frequently visited Universities to examine and criticise proposals. I was responsible for Cambridge, Newcastle, and all the Welsh Universities. I was a member for the next 6 years and I made a significant contribution particularly on Microprocessors.

Chapter 12
Human Computer Interaction (HCI)

My experience in convincing the Computer Board to upgrade the Liverpool KDF9 to an ICL 1906S greatly influenced my views on computing. I realised that the interface between Computer Applications and the User was really important and was a major factor in the implementation of successful computer applications. All readers must have experienced the frustration of trying to use a Web Site which did not seem to understand what they wanted! In 1974, many computer scientists viewed user interaction as unimportant and regarded the mathematics as more relevant. At the time there were a few research workers interested in the subject and it was referred to as Man Machine Interaction (MMI) but this was later changed to HCI (Human Computer Interaction) to eliminate the sexist implications. In the USA, it is called CHI (Computer Human Interaction). In the early days, there was a lot of criticism of HCI research at the time from the more mathematically orientated computer scientists who saw HCI as a distraction from the important issues of Software Design. Today, however, many designers regard it as vital to the success of an application. There is no point in having a good design if it is unusable.

Some readers may not be familiar with the term Human Computer Interaction (HCI) so, to begin with, here is an explanation of what HCI is, and why it is important. At the simplest level, good interface design simply means making computer interfaces easier and more efficient to use by a wide variety of users, from novices to experts. All computer users must have experienced the frustration caused by computer applications which do not understand what the user is trying to do or do not offer the options that the user requires. This usually happens because the designers of the interface have not anticipated this requirement. How many times has the reader been completely stumped by what to do next on a web site?

The reason for this is not hard to find. In the 1960s and 70s designers tended to assume that users had the same approach to the application as themselves, not realising that the human brain is quite complex, and users may have very different ideas and motivations from those of the designer. In the early days, when computer applications were used by a small number of specialists, the approach often met the minimum requirements of most users, but when computing applications, particularly after the introduction of Microtechnology, expanded to include huge numbers of naive users, who were not really interested in computer techniques (users who just wanted to get a job done), such interfaces were found to be woefully inadequate.

Whilst good HCI is a real benefit to many users, there are some application areas where good HCI design is not only desirable but essential, and where errors in the interface can cause serious accidents. Complex interfaces, such as those controlling Nuclear Power Plants and Aircraft, require really well-designed interfaces to avoid serious accidents, and in some cases bad design has actually caused accidents.

In the early days there was also a second reason for poor interface design—a lack of sufficient storage. In the early computers, good interface programs were usually too large for the current main store available and slowed down the interaction too much, but with the availability of more storage, good interface design became possible.

After the Manchester experience, I wrote an early paper for one of the computer journals on the techniques we had used to encouraged users to increase their submission of jobs to Manchester thereby increasing usage of the Regional Centre. I pointed out that the problem was partly caused because of the inadequacy of the User Advisory Service at the Liverpool Computer Centre, and I suggested that Advisory Services in Computer Centres ought to be looked at to see if they could be improved.

I therefore submitted a bid for a research grant to the Science and Engineering Research Council (SERC) late in 1975 for money to employ a Research Assistant to examine how effective current Advisory Services were, and how they could be improved. The response indicates how, at that time, people didn't understand problem. The bid was first turned down by the SERC Computer Committee because they said it was too Psychological and they suggested it would be more sympathetically received by the Psychology Committee (ESRC). I therefore revised the bid and submitted it to the

Psychology Committee but they, in turn, said it was too Computer Science orientated!

I talked with Robert Auld at Birmingham University Computer Centre. He had expressed some interest also in this type of the work and we rewrote the application as a joint research grant from the two University Computing Centres. In the summer of 1976, we eventually obtained a £60,000 grant from the ESRC for two Research Assistants for three years, one at each site to examine the workings of the Advisory Services at Computer Centres.

On 30 October 1976, I interviewed a research student called Mike Coombs for the job of Research Assistant funded by this new grant in the Liverpool Computer Centre. He fitted the requirements perfectly and this turned out to be an important appointment. Mike was a graduate in Psychology (originally from Liverpool) and had just completed his PhD studies at Birmingham. Mike worked with me closely for the next eight years His influence was important because he brought the psychological aspects to our work in HCI.

We commenced our research by studying user attitudes across Faculties. There were big differences. Whilst all Faculties regarded the Advisory Services as important and liked both the courses given and the documentation provided, Advisors from the departments themselves and Computer Centre Staff were rated poorly. Non-Scientific users often spent a great deal of their time trying to solve failures before consulting the Advisory Service so that service was only seen as a final port of call for solving the problems, and 40% of users said that they did not understand the advice given and claimed that the advisory interaction was either painful, or only moderately pleasurable! Interestingly users were quite sympathetic with the Advisor's task and felt it was a difficult job.

We carried out additional studies at Imperial College, and at the Universities of Swansea and Surrey. The overall initial conclusion of the study was that the problems of the Advisory Service were primarily concerned with communication rather than the advisors' ability to solve problems. Social Science, Medicine Arts, users did not share the "scientific culture" of the Advisors

Studying these results, it was obvious how time consuming these Advisory Interactions were, and that one possible solution we proposed would be to incorporate more of advice and guidance *in the computer itself.* In other words, the designers of computer services should spend much more effort in developing Advisory Software which could assist users with computer problems on-line.

However, this only became feasible as main storage expanded and became cheaper in the early 1980s.

Mike Coombs and I realised that User Interface Design was a new and important research area and Mike suggested that we organise a conference early in 1978 at Liverpool for the few people we knew who were currently interested in HCI (Human Computer Interaction). This work enabled us to contact others working in the area such as Edmonds, Shackel and Eason (Loughborough University), De Boulay (Aberdeen) and O'Shea (The Open University), Green, Sime and Fitter (Sheffield), and Thomas (Leeds). The workshop was held at Liverpool in March 1978 and was probably one of the first conferences (if not the first!) to look specifically at HCI issues.

It was a small conference, held at Greenbank House in Liverpool over two days and about 20 people came. This included several researchers, who later became well-known in HCI research, who joined Mike and I to give papers at the conference. Curiously, a person from the academic publisher, Academic Press, heard about our plans and rang up asking if she could come to the conference. She said that she thought there might be a book resulting from the conference. I was surprised and doubted it, but after the conference Mike and I were asked to author a book on the conference proceedings, which turned out to be the first book in that research area—*Computing Skills and the User Interface*.

We took great care with the book, insisting that all papers were of an appropriate format and the design on the front cover became the standard for many subsequent books in the *Computers and People Series* from Academic Press. It took a long time to complete and publish and was released early in 1981. About 2000 copies were produced and it was quite expensive. I thought that they would not sell many, but eventually every copy was sold!

At the conference, Prof. Ernest Edmonds introduced the idea of a front-end dialogue system which provided the required interaction with the user and was separate from the application. This meant that the dialogue could be adjusted depending upon the skill and expertise of the user without reprogramming the application and this would be enhanced by the Dialogue System having access information about the users' habits and requirements. Later the term User

Interface Management system (or UIMS) became widely used for the combined Dialogue System and User Application Model.

At the end of 1979, Mike Coombs suggested that I had better become more familiar with Psychology, so I enrolled for a very successful Open University course in December 1981—A high-level course in Psychology. It was an excellent course and I passed with Distinction (thank goodness because I knew some of the Professors there!). I learned a lot about short and long-term memory, statistics and semantic nets. This knowledge stood me in good stead later on in my research.

Even in the late 70s there was still considerable hostility from commercial programmers and some Computer Scientists towards HCI research. I remember, whilst still in Liverpool, I was asked to give a talk on HCI to a well-known software design firm, holding their annual conference near Gatwick Airport. I faced a lot of hostile questioning. "Of course, we think about users!" they said, "Do you think we don't care?" At the end of the talk, I suggested that my research team went in and examined one of their current on-going designs and they readily agreed.

Four weeks later, we presented our results. They had made many classical errors in the design of this interface. In one case the software went into an endless loop. When we presented the results, the Software team were very embarrassed. As the meeting progressed you could feel the white flag slowly going up the mast and they agreed that HCI design was actually very important!

I also remember going to Cardiff University Computer Science Department as External Examiner in the early 80's where the other external examiner was a "Mathematical" Computer Scientist from Loughborough (who used to work with Alan Turing). I remember him referring to Prof. Ernest Edmonds (who was also at Loughborough) saying "All he does is bloody HCI" (I don't think Turing would have had that view!).

Things began to change when microprocessor systems were introduced, and cheap storage became available in the early 80s. Interfaces had to change dramatically, because of the much wider user-base that micro-processing had created. The new users demanded simple and easy to understand interfaces. The MacIntosh computer interface was specifically designed after much research had been carried out at Xerox Parc on interface design. This brought in the first graphics-based windows system which Microsoft later copied. HCI eventually became a major research area across the world. Now there are large annual conferences – HCI conferences in the states, INTERACT conferences across the world and HCI conferences in the UK.

Chapter 13
Post Codes and Drama at Llanrwst

Whilst we were preparing for installation of the 1906A in 1974, an event happened which was important for data processing but went largely unnoticed by the academic data processing community. This event was the introduction of the Post Code in the UK (originally called The Postal Code). The Post Code was developed to make letter and parcel sorting much  quicker and less error prone. The idea was to encourage people to add the post code of the address to the address on the envelope being posted. An identifier could then be added by the post office (or even better recognised automatically) and this identifier would then be used to automatically route the letter to the correct delivery sack.

The Post Office in the UK had been progressively developing the Post Code since 1959. The initial system of named postal districts, had been developed in London over 150 years ago (around 1860), and this was increasingly extended to other major cities. Various experiments took place in the 1950s and in 1965 Tony Benn (the then Postmaster General) announced that post coding would be extended to the whole country during the next few years.

The Post Office hoped that commercial firms would rapidly adopt the system and that its use would gradually be taken up by individuals. One initial problem with the idea was that people did not know their own post code, or that of the person to whom they were sending letters. In 1970, the general public were therefore reminded to "Remember to use the post code" on their Christmas Mail, though most people had no idea what their post code was. By 1974, the Post Code was fully defined and ready for use but hardly used by anyone!

The format of the Post Code is *Outward Code Space Inward Code*

The Outward code is between two and four characters long. It is usually closely related to the name of a local town or area of London, followed by one or two digits indicating a district in that region. Examples would be L21 (Liverpool District 21), OX28 (Oxfordshire District 28), SE10 (London District SE10).

The Inward Code consists of three characters, an initial number (the sector number) followed by two letters representing a much smaller area within a sector (typically 60—140 addresses). Examples would be 1AD or 7HF. This assisted the Post Office in delivery within a district. A complete Post Code Example is OX28 1AD. The Post Office also allocated mnemonic post codes to highly prolific organisations e.g. BX for Banks.

The Post Code offered huge advantages to Data Processing, but it was somehow ignored by most businesses. I suspect this was partly because of the large cost of post-coding their files which was not a trivial task. A Post Code identifies the location of a UK address to an area usually within about 70 or 80 individual addresses. This means it is really useful for delivery schedules, planning routes, and even characterising social areas. However, in 1974 most data processing firms did not seem to have realised the Post Code's importance.

At the University of Liverpool Computer Centre, we came into contact with Post Codes by a happy accident. In November 1994, a Mr Armstrong of the Merseyside and North Wales Brewing Association telephoned me at the University. He said he had read our advertisement in the Times and would like to talk about a possible contract for some computing work. In fact, we had not put an advertisement in the Times, but I didn't admit that and asked him for more details of his requirement.

He pointed out that in Wales they voted every seven years to decide if the pubs should open on Sundays (known as being "Wet" or "Dry"!) and that the election was coming up again in November 1975. At that time nearly all of the counties were "Dry" on Sundays. For a county to have a vote on the issue there had to be a requisition supported by at least 500 voters. The Merseyside Brewing Association (along with their South Wales counterpart) would be requisitioning all the Dry areas and, no doubt, the Church would requisition all the Wet areas. This would mean an election involving more than 2 million voters.

The Brewing Association was convinced that a high poll would be an advantage to them and the way of ensuring this was to issue poll cards to all voters with the address of the polling station on them. However, the Government were not prepared to do this for this type of election. At the previous election in 1968 the brewers had tried to issue poll cards, but it had been a complete failure. Mr Armstrong wondered if we could do a better job! He pointed out that the actual polling station locations would only be known 2 weeks in advance.

I replied that I thought we could do it and that I would get back to him. I could see that we would have to capture the addresses of the two million voters in Wales on the computer, and we had nearly a year to do that. But what worried me was how we would match up the polling station address with the voters addresses, having less than 2 weeks to do it, not only to complete the match, but also to deliver 2 million poll cards on time.

We then had a stroke of luck. I don't remember how it happened, but I got in contact with a Mr Terry Lipson of O.E. McIntyre Ltd. which was a Mail Order firm based in Liverpool and who were used to handling large quantities of mail. Terry was a fascinating person and we agreed to meet him. He thought that between us we could handle the problem because he routinely handled millions of addresses across the UK. His stroke of genius was to suggest that we use the post code (just announced by the Post Office the previous year) for matching the stations with the voters' addresses. It sounded exciting, but I asked Terry, how on earth could we add the Post Codes to 2 million addresses?

He replied that the Post Office had a file which contained about 80% of UK addresses already post coded. For the other 20% he claimed he had the capability at O.E. McIntyre to add the post codes to the rest of the addresses manually, and then these could be added to the 80% to form a complete post code database for Wales. Provided we received the Post Codes for the polling stations, matching would be easy.

I returned to the brewers, and they accepted our proposal. Terry Lipson, my Applications Manager (David Goddard), myself and a new programmer, who had just joined my staff at the Centre, called Bob Carter, formed a Partnership called "The Merseyside Computing Association" to handle this business. Bob Carter was a first-rate programmer and I never ceased to marvel at his competence. He often used to say that there were only five computer programmes in the world and that all he did was alter them to suit his purposes! He never explained what those five programmes were, but I guess he was right. Throughout the first half of 1975 we created our own files from the Post Office data and the Terry added the manually captured data.

By September 1975 we had captured all of the electoral roll of Wales and stored it on the 1906S with the required post codes. The polling station data arrived, and we printed all the poll cards on the 1906S fast printers and shipped them to O.E. McIntyre Ltd. Terry and his staff then dispatched them. Early in October, the Brewers reported that virtually all of the poll cards had been successfully delivered (in contrast to 1968) and polling day (November 5th) approached. The Liverpool Daily Post announced our efforts (see below).

The evening before polling, BBC TV decided to do a feature on the Welsh Sunday opening. They based themselves at a pub in Llanrwst and I went there to take part in the broadcast. The Brewers had asked me to predict how the polling would go and I had produced a report suggesting that most of Wales would switch to drinking alcohol on Sundays. They wanted me to push this idea on the programme. The result was a famous TV programme which was remembered for years afterwards! The programme went down in the BBC history archives as one of TVs incredible events.

I went down to the pub in Llanrwst on November 4th, to take part in the live programme with representatives of the Brewers, the church, and two "ordinary people" from North Wales. The link man for the proceedings was Dennis Tuohy, a presenter on the Tonight program which in 1975 was a very popular TV programme (often with presenter Cliff Michelmore). The program (which was LIVE) was scheduled to go out at 10pm from Llanrwst, but that night General Franco, not always a reasonable man, had a stroke and the BBC decided to lead with a program about him first.

We were all in this tiny bar in Llanrwst. There was a roaring fire. Dennis Tuohy was in an armchair, and it was so hot that he had two make-up artists behind his chair so that when the camera was not pointing at him, they powdered his face. Nobody powdered our faces! We sweltered in the heat and waited for the Franco program to finish. We were all sat round a low table opposite Dennis Tuohy and the table was ringed with microphones. The producer, in a flowery shirt, was behind the bar with the cameras.

At about 10.15pm we went live. I was in the middle, flanked on the right with the Brewers, and on the left, was a lady from Aberconwy ("Mrs Evans" representing the ordinary voter). Dennis opened with a general explanation of what the election was about and then asked Mrs Evans to say in a few words what she thought about Sunday opening. She began speaking and I noticed that she seemed to be holding a large script. The vicars on the far side of her were murmuring approval in Welsh, as she spoke against Sunday opening. After about 3 minutes Dennis Tuohy thanked her for her contribution and tried to bring me in. However, she ignored him and carried on speaking.

After another 30 seconds Dennis Tuohy made a desperate plea for her to stop but she completely ignored him again and carried on. I could see that she still had many pages of script to read! Dennis then introduced me over her soliloquy and asked me to comment on my computer predictions for the outcome of the vote. She never stopped talking and I had to carry on speaking loudly above her! The Producer had given me McKenzie's Swingometer (used in TV coverage of National Elections) and I had it on my knee. Dennis, who was still trying to over-talk Mrs Evans, tried to make a joke about me using a "Swigometer", but it was lost in the confusion!

I desperately tried to carry on. By this time, Mrs Evans was throwing her arms about and looked as if she was going to hit me. In fact, my wife who was watching the program thought she had hit me! Now, even the vicars were telling her to be quiet but to no avail. As I started to talk about the predicted swing of the votes, a hand suddenly grabbed my leg, I let out a cry and dropped the Swingometer. There was an engineer under the table trying to disconnect her microphone but he grabbed my leg by mistake! Total pandemonium ensued, and this was all live, but they could not cut her off. Dennis bravely carried on and eventually we went off the air with her still speaking!

Dennis put his head in his hands and said, "I am finished as a presenter." The producer was wringing his hands, was almost in tears and rushed off to phone the BBC. Mrs Evans seemed completely unaware of the chaos she had caused. Then suddenly, the producer rushed back into the bar. "Wonderful" he said

"Wonderful—all the phone lines at the BBC are completely jammed with people ringing in complaining about "That Woman". Unfortunately, I don't have a copy of the program (I don't think they took copies in those days).

The next day (November 5th) we all went to Llandrindod Wells to cover the voting results. These were also on live television, and I spoke both in English and Welsh. My predictions were correct. There was a very heavy swing to "WET" and many districts voted to open on a Sunday and only four remained dry. It was reckoned that "that woman" had had a significant effect on the vote. For days afterwards, I received letters from people across the country (who knew me) commenting on the program and mentioning "that woman".

The following month the Brewers congratulated us on a brilliant job and we each received two dozen bottles of fine champagne. The real hero, however, was Terry Lipson who had also done as great job in delivering the poll cards.

After the Wales election I wrote a paper on our experiences for the Brewers. Terry Lipson became quite excited at working with us and we realised that our work could be exploited. We decided that Post Codes were much more important than a technique for delivering letters. A Post Code identified an address usually to within about 100 yards and could be used for all sorts of applications. The problem, however, was still the lack of Post Codes in the addresses of commercial files in the UK—to fill these would be a mammoth task. Terry suggested that if we could persuade two firms to join with us in post coding their files, this would provide a complete database of Post Codes for the UK that we could then sell on to other firms. Two applications areas we had immediately in mind were Address Duplication and Dispatch Scheduling.

Address Duplication was an important application. Two of Terry's interested customers were Readers Digest and General Universal Stores (GUS), both based in Liverpool. Terry informed us that most of the second-class promotional post was posted in Liverpool because of this. Both Readers Digest and GUS were concerned because they had people placing orders for postal delivered goods, and some of these customers were fraudsters. These people would order, say about £100 worth of goods, and then not pay for them. Since the amount was quite small, Readers Digest and GUS would write them off and put a stop on their account. But the same people would then order again. The trick they often used, was to order again from a nearby address and since postal delivery workers are often able to correctly deliver wrongly addressed letters, the goods could be successfully delivered to a completely different address. The fraudsters would register the new address and more goods could be delivered there (again not paid for!). Terry pointed out to Readers Digest and GUS, that if they post coded their files, the Post Code could be used to match up nearby addresses and catch out the thieves.

To check the effectiveness of posties delivering letters, I decided to carry out an experiment. I sent about 30 letters to myself, each with an error in the address. The street might be spelled badly, or the wrong number for the house, or my name might be spelled wrongly. Astonishingly, the Postie managed to deliver 90% of the letters (the few which were not delivered were because of a serious error in the surname). We realised that this was why people could order more goods from Readers Digest. They would put the correct name, but with a house number a few doors away and the postie would still deliver a letter to them. However, using the Post Code, we would be able to check the names of people in the 70 or so adjacent houses.

Our system had worked well for the Welsh Sunday opening using the 80% correct files from the Post Office and then manually adding the other 20% using the staff of O.E. Macintyre. So, for Readers Digest and GUS we would create the complete UK Post Code Data Base, by the same approach. It was a big job, but we felt that once we had the database, we could offer to Post Code other firms files using our system. Readers Digest and GUS expressed keen interest and we went ahead.

This meant first setting up a data file of all the postcodes in the UK and we used the current Post Code File of the Post Office which covered 80% of UK addresses. Bob Carter wrote programs to convert the 80% UK Post Office File to a file format acceptable for Readers Digest and GUS, and O.E. McIntyre Ltd agreed to hand-code the difficult addresses. This was a big job but eventually we had a complete UK Post Code file.

This convinced Readers Digest, and we converted their address files to address files with Post Codes added. The new de-duplication system worked really well with impressive results, so we then used the file to Post Code the GUS files.

Bob Carter then wrote a modification to our computer programmes so that we could input the address formats of different industrial concerns (these vary a lot) and then use our program to automatically add post codes to their address files. I remember Terry employing someone to sell this post code conversion system to customers. This person had little idea what the system did but still managed to successfully obtain orders for a large number of post code conversions! We took a standard fee for conversion. These days of course everyone uses software like ours to identify peoples addresses. It asks you for your postcode and then ask you to choose the address number. If only we had patented it and charged a halfpenny an access, we would be millionaires! Later Bob left us and joined Terry at O.E. McIntyre Ltd.

Post coding was a major activity for us over the next two years. I wrote two reports on post coding and a promotional booklet called "Postcode for Profit". We also developed a sales programme based on post coding called Salesmatch, for scheduling and marketing done in collaboration with O.E. McIntyre Ltd. Today, of course post codes are used everywhere (for example in satellite navigation systems and in dispatching) but I think Terry and ourselves saw the opportunity early on.

Was this one of the first occasions when a private consortium used a computer to influence a political change within a democratic state? Now, "Big Data" has become commonplace.

Chapter 14
Databases and Visits to Cairo

In 1978, with disk space becoming cheaper, we became involved in Database developments and installed IDMS (Integrated Data Management System) on the Liverpool 1906S. This was a very powerful Data Base Management System (DBMS) and offered significant improvements in handling and accessing data.

A DBMS allows the user to create and manage data in a secure and highly organised way. In the early days of computing programmers would write an application and the data required would be created in a file. Then other applications were written, and different files created. Often data was duplicated across files. As data processing became more complex it was soon realised that many non-connected files were being created. For example, take a local doctors surgery which has a file of all the patients, and their home address, illnesses and treatments etc., doctors might want to examine the data from different points of view. They might want to use the data of a whole family, or examine all cases of flu. The data will have lots of cross-links and a database allows the data to be stored with these links in place. So, rather than many non-interconnected files, there will be one large base of data with the links between different data items.

The DBMS acts as an interface between the data and the application. It provides an engine to organise and protect the data, and a Schema (or plan of the data) which defines its logical structure (i.e. all the links). It connects together, in a hierarchy, all the relevant information about the people addresses, functions etc. A DBMS will also have powerful back-up and recovery mechanisms as well as logging, roll-back and auditing functions. It provides many different views of the data for different users and the users need not know where the data is stored. Many of the research workers at the University found the DBMS approach useful as well.

In November 1979, ICL (Israel Chemical Ltd) heard about our Network and Database work and approached me and asked if I would give a talk at a seminar in Cairo. ICL had installed a 1906S there, like ours, in the Statistics Department at Cairo and the head of the Unit—General Askar—was holding a 4-day conference in January 1978 in Cairo. I agreed to contribute, and was told that Jack Howlett, David Firnberg (Head of the National Computing Centre), and others were also participating, and we could bring our wives. ICL also expected that we, at Liverpool, would be approached by the Egyptian Government to offer assistance in installing IDMS on the system in Cairo.

The Cairo conference took place on 14 January 1980. ICL booked us all on Kuwaiti Airlines to Cairo. About four days before the flight, ICL rang me to say that the flight would not have any alcohol on it, but that Kuwaiti Airlines had offered to put some on for us. What sort of drinks would we like? I (without thinking) suggested 12 bottles of red and white wine, and several bottles of whisky and gin. When we arrived at the Terminal, we could see the Airline hoisting the booze onto the plane! We had a great flight. Because the plane was flying to Kuwait, the airline was only allowed a limited number of passengers to Cairo, so the plane was half empty. There was a lounge at the rear of Second class which we occupied and had our drinks with soft drinks supplied by the Airline. On the return journey we continued to drink the rest of the supply. I remember passing down the aircraft and a seasoned passenger beckoned me. "I am an oil worker and travel regularly on this flight. They don't have alcohol on this plane" he said "but I bring my own drinks. Come and have one with me!" I did so, but then I told him about our supplied drinks by the airline. He nearly burst into tears!

We stayed at a beautiful hotel (The El Salaam Hyatt) and had taxis each day to the conference. On the second day I gave a presentation on "University Networks". I remember General Askar, asking David Firnberg to sum up the conference in the final morning session. This was a one-hour slot and David had to work most of the evening to prepare his one-hour speech. The next morning, as we walked in, General Askar handed a sheet of paper to David and said, "this is what you should say!" It consisted of about half a page which took less than five minutes to deliver. The conference was over by 9.45am!

After the conference, we all went to Luxor for a few days to the Valley of the Kings. This was a magnificent trip. We visited the Tombs of Tutankhamun, Rameses and Hatshepsut and the magnificent temple at Luxor. This was probably the finest visit of my life. We also saw the first Djoser Pyramid at Saqqara, and after that visited the Colossi at Memnon. These are two gigantic statues by themselves on the edge of the dessert, each weighing over 500 tons and over 4 metres high. The mathematician Jack Howlett stood before them and recited the famous poem Ozymandias by Shelley.

I met a traveller from an antique land,
Who said: "Two vast and trunkless legs of stone
Stand in the desert. Near them, on the sand,
Half sunk a shattered visage lies, whose frown,
And wrinkled lip, and sneer of cold command,
Tell that its sculptor well those passions read
Which yet survive, stamped on these lifeless things,
The hand that mocked them, and the heart that fed;
And on the pedestal, these words appear:
My name is Ozymandias, King of Kings;
Look on my Works, ye Mighty, and despair!
Nothing beside remains. Round the decay
Of that colossal Wreck, boundless and bare
The lone and level sands stretch far away."

After the conference, ICL negotiated a contract for us (at Liverpool) to work with General Askar in developing data bases by installing the software IDMS (which we were already expert in using). On 24 November 1980, I went to Cairo with John Martin (our Applications Manager) for discussions with the Egyptians on further developing IDMS. We were there about 10 days. We got on well with the Egyptians, successfully installed IDMS and planned future visits. The Egyptians looked after us well but couldn't pay us very much. I therefore suggested that, instead of paying us money, they should fund myself and the team to visit places of antiquity each weekend and they indicated they were more than happy to do this.

On one of the intermediate weekends, they took us again to the Valley of the Kings. It was a wonderful visit. On the Saturday afternoon I went swimming in the Winter Palace Hotel at Luxor and the boys round the pool kept saying "You famous Film Star—give us some baksheesh" (baksheesh means a bribe or a present!). I knew I was good looking (!) but didn't understand why they were saying this. In the evening we went into the Winter Palace for dinner. It was a buffet with lots of curry dishes etc. As I helped myself to curry, I saw Tom Baker (see right, who was the fourth TV Dr Who, 1974—1981) looking at me and I realised that I probably looked a bit like him. I said, "you have caused me a lot of trouble giving out Baksheesh to the boys because they thought I was you!"

"Oh dear," he said, "I'd better buy you a drink. Come and join us." I went over to his table, and he was sitting with Raymond Burr (some may remember him as Ironside and Perry Mason on television) We had a good evening's drinking late into the night! They were there to make a programme for BBC Wales. I later saw it, but I thought it wasn't very good!

We continued travelling to Cairo showing the Egyptians how to exploit IDMS for the next year. On one occasion I met the wife of the President of Egypt, Anwar Sadat.

They also took us to the Aswan Dam, and to Abu Simbel. We enjoyed these visits and IDMS worked well. In September 1981, I was invited to London to have dinner with General Askar, in London with ICL This was the last time I met him. He died shortly afterwards of cancer. Then, on October 6th the President of Egypt, Anwar Sadat, was assassinated and our relationship with Egypt came to a sudden end.

Chapter 15
The Microtechnology Revolution: The 'Alty' Report

Whilst working on postcodes early in 1976, I noticed snippets beginning to appear in the press about what were called Microprocessors. The comments seemed to imply that a revolution in computing was about to happen. There was talk of mainframe computers becoming obsolete and I took them seriously, partly because I perceived that they might be a threat to my position as Director of the Computer Centre! It was obvious that university researchers would become involved with these devices soon and they would need assistance, particularly in the early stages of development. I decided to find out something about them. What I found out made me realise that something very important was taking place and it would change the nature of computing provision.

Firstly, it rapidly became clear that their internal organisation was not significantly different from the standard computer. Indeed, they were just Turing Machines. Alan Turing would have regarded them as a natural extension of what he had proposed. They contained all the usual things you normally found in traditional computers. The only difference was that the rapid development of the technology had allowed more and more circuits to be accommodated in less and less space. This meant that computers would get smaller and smaller and require less and less power. This would also mean that computers would spread into every device large or small and the nature of computing would totally change.

Fortunately, one of my staff had just left so I had a vacancy. I immediately advertised for an electronics expert who knew something about the new technology, and I was very fortunate in appointing Dr Malcolm Taylor, an Electrical Engineer, to assist us. He was a real  asset and strongly influenced our thinking. Malcolm quickly got hold of a couple of microprocessors (an Intel 8080 (right) and a Motorola 6800) and played with them. At the time, we were hearing very little about what other Universities were doing in this area, so we organised the first Liverpool Microprocessor Workshop in September 1976 (almost certainly the first Microprocessor Workshop held in the UK) and invited most universities to generate interest.

About 50 people attended, who were mainly research students, since most of the more senior staff new nothing about Microtechnology! The conference was held for two days at Greenbank House, a nice location in South Liverpool owned by the University and two days of research papers were given. At the end of the first conference the delegates felt that the conference had been a great success and that there should be a regular communication mechanism between interested workers in the field. Malcolm and I therefore created a sort of informal Journal called MICROSWAP which, for a few years, was the main newsletter on Micro activity in the UK. Later this was converted into a formal journal (The Journal of Microcomputer Applications) with Malcolm and I as editors and published by Academic Press. The Liverpool workshops were very popular, and they became an annual fixture. We continued to hold them every year until 1982.

Once we understood what Microprocessors were about, we were able talk to researchers in the various Liverpool University departments who were beginning to purchase small systems. A policy was needed for supporting them and the Computer Centre was the ideal place from which to do this. The initial idea was to define which microprocessors the Centre should support, and to discourage users from buying others, but Malcolm argued that the development needs were so fluid that it would be best to simply go with the flow. Users were told that the Centre would try to support whatever they bought but that the current major expertise of the Centre was with the Intel 8080 and the Motorola 6800. This policy worked well because it meant that users mostly bought the microprocessors that the Centre could support, but importantly no other microprocessor was excluded.

Every spare bit of cash I had I ploughed into the formation of a Microtechnology Support Unit. It was quickly realised that users would need software development systems in which to create the software. With the cash, a medium-sized Microprocessor Support Laboratory was quickly built up at Liverpool in 1976 which Malcolm ran. It was very successful, and Microtechnology rapidly took off across the University and local industry also became involved. Within one year, nearly £75,000 had been spent in building up this facility. The Liverpool Workshops, which were being held each September became very popular and knowledge spread rapidly.

In 1977, the Computer Board, on which I was a member, had also begun to worry about how such processors might affect their approach to the provision of computing power at Universities and Research Councils. The Department of Education and Science had produced a report on computing in Education about two years earlier and astonishingly had not even mentioned microprocessors! The Computer Board therefore asked me to form a working party to examine the situation and report back on how Microtechnology might affect the provision of large computers in universities. I asked an impressive group of people to join the working party—Iain Barron (of INMOS), Prof David Aspinall (of Manchester who was at school with me!), Prof Basil Zacharov (whom I had known when at Daresbury, who had later become Director of the London Computer Centre), and Mervyn Williams of the Rutherford Laboratory. This was a small but high-powered body and had its first meeting on 3 October 1977.

The working party worked well together, and the report was completed within four months. The contributions from every member were outstanding. The report was presented to the Computer Board in early 1998 and was known as the "Alty Report". The Board received it enthusiastically.

Previously, the general opinion in the computing world at that time had been that there would be a rapid shift to computing being spread over the whole University with each department, and even individual members of an organisation, having their own powerful computing system on their desks for a few pounds. There would be no need for central facilities and eventually mainframe computers would become obsolete. That didn't sound too good to me being the Director of a large centre!

Whilst the working party agreed that there would be a dispersal of computer power, they pointed out that far more fundamental consequences would result from the reduction in size and huge increase in storage capacity. These two factors would fundamentally change the nature of computing. Soon computing devices would be found in almost all equipment—radios, washing machines, cookers, and virtually everyone would be in contact with computers even if they did not realise it.

Thus, this expansion into new computing areas, and the introduction of computers to wide sections of the community would dramatically widen their use. The new computers would be very similar to existing computers in that they would consist of hardware and software but the ratio of costs between hardware and software would dramatically change. Whereas the hardware costs would continue to fall, software costs would continue to rise and so the overall cost of providing large scale computing would not fall as dramatically as expected (no-one previously had thought about this aspect). The working party therefore identified two distinct areas of development—High Volume/Low-Cost and Low Volume/High-Cost categories.

The former category (High Volume/Low-Cost) was expected to produce millions of chips and would be very cheap indeed. These would often be single chips and probably would not be even recognisable as computers. They would be found in everyday devices such as washing machines, cookers, TV sets etc. High sales volumes would be required to keep costs down and new applications areas would be essential since the current data processing market could not provide a viable marketplace for such chips. The Report estimated that the data processing capacity of the world in 1978 represented 2 weeks' production capacity for the chip manufacturers!

The latter category (Low Volume/High-Cost) would almost certainly result in new special purpose devices such as Distributed Array Processors, Data Base Engines and High-Performance Interactive Computer Systems. For example, the Working Party identified that mass storage of data would be very economical, and this would result in large databanks accessed by huge networks. Remember that in 1978 the internet had not been thought of! The Working Party therefore expected a vast range of computers to become available from small personal computers costing a few hundred pounds to very large systems costing a million pounds connected by networks.

Software for the new technology would also be required in both categories. The software in the single chips would be single purpose only, very cheap if only because of the high volume of sales. It would be unalterable by the user and may not be maintained – it would *simply have to work*! Because computers in the future would be used by most of the population, not only will the software have to work, but it will also have to be tolerant of human frailty, and interface development will need to be more sympathetic to user needs.

In the Low Volume/High-Cost category whereas the hardware would be very cheap the software costs would increase and would be significant and this would demand maintenance which would increase costs further. Thus, the end user will see falling hardware costs but less stable and rising software costs. Software costs would also dominate the High Volume/Low-Cost applications.

Because of this the working party warned users about "going it alone". It warned that software development would require debugging systems—logic state analysers, simulators, emulators, and development systems and recommended that Micro Support Laboratories (based on the Liverpool Model) should be introduced into University Computer Support Centres. This would not only promote the introduction of Microtechnology into University Departments but would accelerate its take-up in industry. It also foresaw rapid developments into 16- and 32-bit microprocessors.

File store support and network developments were regarded as key aspects of the new environment. Files were the basic material of any data processing system and that the developments in micro technology had already rendered floppy disks and bubble stores obsolete. It estimated that the billion-byte disk was only a 2 to 3-year time away, and a computer centre with huge disk storage would be able to maintain a large collection of files cheaper and more safely than dispersed storage systems. As a result, networking would become a critical feature of the new environment,

The report was well received and immediately adopted at the Computer Board. When I presented the report at the Board, a representative of the Treasury was sat next to me. He quietly said to me "If you can get a case to me today, I will get you £2 million to implement the first part!" I replied, "Do you mean now?" He replied "Yes." So, whilst the Board Meeting continued, I scribbled (in pencil) a one-page bid and gave it to him for £2 million. The next week the Computer Board received the £2 million to implement the initial proposals to set up other micro laboratories at universities following the Liverpool Model.

The report was well received everywhere. The Times called it "the most important report since the Flowers Report." A colleague of mine, Prof Jim Zissos, had recently written a book with the title "System Design with Microprocessors". He asked me to write a *Foreword* for it. When the book was reviewed, the reviewer commented "The *Foreword* by Prof James Alty is one of the most sensible comments I have ever read on Microtechnology. It should be tattooed on the Department of Industry's backside!" That made all the work worthwhile!

Shortly afterwards, University Computer Science departments complained that they should have had the money (not the Computer Centres)! The Chairman of the University Grants Committee (which funds the main University departments) was concerned by the report, and he said he might try to block it because it only gave money to Computer Centres. I met with him in front of the Minister for Education (Rhodes Boyson). I pointed out to him that, rather than oppose it, he should welcome it and use it as the basis for an even bigger grant application for Computer Science Departments. He agreed and eventually obtained a further few million pounds for Computer Science departments! During the next two years, I spent much of my time lecturing at universities around the country pushing the establishment of the Microprocessor Laboratories, holding business seminars, lecturing to large industrial concerns and encouraging industry to adopt Microtechnology.

I think that the working party did a brilliant job, and I hope that it was partly responsible for encouraging industry to take up the challenge. The Working Party was fortunate in being a collection of really bright people.

The new Microtechnology had a dramatic effect on computing with new start-up companies springing up everywhere. In 1975 Microsoft was formed by Bill Gates to sell BASIC interpreters for Altair. It went live in November 1975, and it was about this time that the company name *Microsoft* was first used. By the end of 1978, Microsoft was turning over $1 million dollars.

Meanwhile in 1976, Steve Jobs and Steve Wozniak founded Apple. The Apple 1 personal computer went on sale in 1976. The price was $666.66. Although not that cheap for an individual, it was an astonishing reduction in price for an individual computer. Apple II (see on right) was introduced in 1977 and this computer had a Floppy Disk Drive. The introduction of the spreadsheet program VisiCalc changed the fortunes of Apple. Revenue grew  exponentially, doubling every four months. Apple III was introduced. In 1980 Apple went public. I think Alan Turing would have been very excited about these developments.

A visit to Xerox Parc convinced Steve Jobs that the future lay in Graphical Interfaces and in 1984 the MacIntosh was announced (see right). This was also an innovation. Good HCI (Human Computer Interaction) appeared for the first time on a small personal computer. The easy-to-use interface was based on solid research at Xerox Parc. It is also claimed that the Desktop Publishing Business was created by the MacIntosh.

The microprocessor report also had an immediate effect on my position at Liverpool. Whilst I had a Professorial appointment at Liverpool, I did not have the Professorial title. I felt that this sometimes put me at a disadvantage in the University. Since we were doing an increasing amount of academic work at the Centre, I thought an academic title would be more appropriate. (I'm so humble!) On 14 January 1979, I applied for a Professorship of Computer Science at Stirling University. However, the tactic here was not to go to Stirling, but rather to put pressure on Liverpool to give me a Chair. The Director at Salford University had just been appointed to a Chair in Birmingham, so it wasn't a unique idea. At the Stirling interview, the Vice Chancellor said that I was "too big for the job!" because Stirling was quite a small University. I agreed with him and withdrew from the contest. I then informed Prof Collinge (the Chair of the Liverpool Computing Committee) in February, about the interview and he agreed to press for a Professorship at Liverpool.

Coincidentally at the same time, Whitbread (the Brewery) were looking for a new Director of IT. They approached me, and I formally applied for the post of Director of IT at Whitbread. This was a top-level post with a high salary, private health care, private education for the children and a move to London.

Whitbread eventually offered me the job and I went to see the Vice-Chancellor, pointing out that I had been offered the Whitbread job and asked what the chances were for the Professorship. He said, "Don't be hasty with your commercial friends." I replied to Whitbread saying I was very interested but that the University was about to offer me a Professorship (something which many friends in the University had been pushing for some time, and I didn't want to let them down). However, if the Chair failed then I would join Whitbread. They agreed to wait.

On July 4th, the Senate Meeting was held where my Professorship was being proposed. I was asked to stay away. The motion was put, and the vote was 97 to 2 (opposed apparently by Computer Science and an English Professor!). So, the next day I was Professor Alty (which meant something in those days!). They had made me the Professor of the Computer Centre which was quite unusual and after I left in 1982 the Chair was never filled again.

The introduction of Microtechnology had a rapid effect on the provision of computing facilities (particularly interactive facilities) as the working party had predicted. At the beginning of 1979 the Liverpool Computer Committee felt that we needed to upgrade our interactive facilities on the 1906S. Microtechnology was already having a huge impact in the development of main line computers and a new generation of high-performance interactive computers were becoming available. Whilst the 1906S was providing quite good interaction, new computers were now coming on the market with huge increases in disk space and interactive capability. Liverpool therefore set up a working party to investigate which computer would be most appropriate.

In March, I went to Newcastle for a meeting as their Computer Board representative and, whilst I was there, IBM announced a new interactive computer—the IBM 4341. It seemed to me that this was an ideal candidate for our interactive system at Liverpool (see right). However, the problem with IBM systems was their long delivery schedule and there was no way a customer could hurry an order further 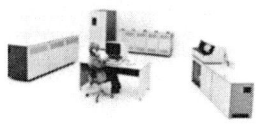 down in the queue. I knew it would be a very popular machine. I talked with Elizabeth Barraclough (the Director at Newcastle) and she said, "Why not order one now?"

"How can I?" I replied, "I must have the approval of the Computer Committee, the Senate at Liverpool and the National Computer Board, and it is possible that other firms might make a better offer."

The IBM salesman was also there. He said, "You can order it confidentially and put a condition on the order that final confirmation needs the approval of your Computer Committee, the Senate and the Computer Board." This I then did, and the North-West Confidential 3 order was born!

Deliberately, I told no one at Liverpool (there was a serious possibility that we might order a DEC system). The University working party met several times. Computers from ICL, IBM, Prime, and DEC were all in the frame. Over the next few months, the committee went to see and evaluate the different interactive systems. The more I saw, the more I was convinced that my idea of an IBM 4341 was the best one.

Finally, the Committee recommended the IBM 4341 and the Computer Board accepted it. The University was slightly astonished when the computer arrived within a few months! A year later I decided to confess to the Chair of the Computer Committee (who was now a Prof Grimley) about what I had done. I admitted to him that I had ordered the 4341 even before the interactive working party had met! I did explain the various caveats I had put on the order. His reply was "Well done Jim, I would have expected no less. That's what we pay you for!" The computer was installed shortly afterwards, and it turned out to be a very successful interactive machine.

People may think that the IBM 4341 was always my first choice and that I manoeuvred the process to get it installed, but this was not true. At the time, I regarded the DEC system as a serious alternative. The caveats I placed on the order could have easily caused it to be rejected.

Chapter 16
Micros Come of Age;
I Go to Strathclyde University

My term at the Computer Board had come to an end (6 years) and I received a letter from Sir Keith Joseph (Secretary of State for Education and Science) thanking me for my efforts and for serving an extra year (the normal term was 5 years). The previous six years had been interesting and I had perhaps had a slight effect on computing in the UK (hopefully positively!).

The Commodore PET (Personal Electronic Transactor) 4032 personal computer (see right) was brought out in 1980 and was adopted by many schools and colleges. Prof Andrew Colin used them at Strathclyde to change undergraduate computing. He set up laboratories of 30 PET computers for undergraduates to use the BASIC Programming Language. Then in 1982, Clive Sinclair released the ZX  Spectrum. It had a colour display and could be displayed on a television screen. It was the first truly mass-market computer in the UK, and sold over 5 million world-wide. In a matter of 7 years the whole nature of computing had changed, and the average homeowner could have their own personal computer (PC). The Spectrum also caused a huge boom in software and games.

Microcomputing came of age when the IBM personal PC was introduced in 1981 and that changed everything. The IBM PC (see right) represented a new platform which—like all changes in this industry—offered a major paradigm shift in computing resulting in many new products with new capabilities. It also introduced  the new era of personal computing. Interestingly, IBM did not realise the power of what it had produced. IBM thinking was still embedded in a main frame philosophy. So, instead of developing the operating system on the PC to a much higher level, IBM let Bill Gates install the operating system MSDOS on the PC, Microsoft was born, and the rest is history!

In the 1980's, the IBM personal computers were coupled with networking and access to powerful databases, with PCs connected to mainframes accessing large data stores (as had been predicted by the working party report in 1978). It also had another desirable effect. With so many more users interacting with computers, the human interfaces had to be improved and the result was the MacIntosh and Windows.

In 1984, Microsoft introduced Windows—the graphical interface system to compete with the MacIntosh. Frankly, Windows is not as well designed as that produced by Apple. The Apple system was carefully designed from scientific research at Xerox Parc, whereas Windows tried to mimic this but introduced some features which did not fit in well with the original design. However, Windows became phenomenally successful. Apparently 4 billionaires and 12,000 millionaires were created from Microsoft employees.

The new technology also redefined the large computer marketplace. In the 1970s there were a number of suppliers of large-scale computer systems in the USA—IBM, DEC, SPERRY UNIVAC, NCR, HONEYWELL, CONTROL DATA (CDC), and BURROUGHS (which merged with SPERRY UNIVAC). Additionally, there were other manufacturers in the rest of the world—ICL, GEC, Fujitsu etc. By the twenty-first century most of these firms had disappeared or changed their focus away from large systems. Only IBM, Fujitsu, and CDC remained the main players in the high-end mainframe business.

In November 1981, the British Computer Society (BCS) was approached by the BBC to explore the possibility of the Christmas television Lectures being based on computing. I suspect the BBC was influenced by the huge effect Microtechnology was having on computing and because of the Microtechnology Report, BCS suggested me as the lecturer. I went down to London in December and gave two specimen lectures to school children at the Royal Institution and the President of the Royal Society was present (Prof Porter).

It is an interesting place. The lecturer is first placed in Michael Faraday's (Chemist, and inventor of the electric motor. Transformer and generator) office across the corridor from the Lecture Theatre. At the commencement of the lecture, the lecturer walks across the adjoining corridor with two laboratory assistants on either side to prevent the lecturer from running away (which apparently happened in the 19th century!). The two lectures I gave went down quite well, but the BBC thought that Computing was, at that stage of its development, not visual enough to be the subject of the lectures so they were shelved. I suspect they were correct.

About this time, the Van De Graaf Accelerator on which I had done all my PhD work was finally closed down at Liverpool and we all had a "wake" party. It had run for nearly 20 years. All this gave me itchy feet. I began to wonder where I should go next. I decided that it was time to move from Liverpool. I therefore applied to Strathclyde University where I knew Prof Andrew Colin, Douglas McGregor and Andrew McGettrick (all of whom had all been active members in the Liverpool Microcomputer workshops). I was interviewed and appointed to the Chair of Computer Science in May 1982 and took up my appointment in October. We held the final Microcomputer Workshop in Liverpool on Sept 6-7 of that year.

My new post as Professor of Computer Science at the University of Strathclyde was quite a change from Liverpool. One reason for moving to Strathclyde was the opportunity of becoming a full-time Computer Science Professor doing mainly academic research and teaching (a professor's main job). At Liverpool, my teaching and research had been more limited although I had published several research papers with Mike Coombs on Human Computer Interaction research.

I always remember going to work on the final Thursday at Liverpool. I suspect my Secretary, Joan, expected me to quietly disappear early, but I worked normally all day as if nothing was going to happen. Then, at 5.30pm, I packed up, said goodbye and went home. On Friday, 1 October, I had a leaving party in the Laboratory together with Mike Coombs, who was coming with me to Glasgow. It was the end of ten years at Liverpool. I had transformed the Computer Centre into an important Centre, I had been one of the pioneers in the introduction of Microcomputing and early research into Human Computer Interaction and we had really pushed the post code. I had enjoyed being at Liverpool, but it was time to move on.

Chapter 17
Strathclyde: Crisis at the Scottish HCI Centre

I arrived at Strathclyde University in October 1982 and Mike Coombs moved up with me. I tried to get him a Lectureship, but the Department were not convinced that he was a Computer Scientist, so they gave him a Research Fellowship instead.

Strathclyde is an interesting University. It is the third largest University in Scotland. Many people will think of Strathclyde as one of the new Universities created out of the technical colleges after the Robbins Report (report of the Committee on Higher Education, chaired by Lord Robbins which recommended the expansion of Universities), in 1963, but that is incorrect. It began as Andersonian University in 1796 and had a Physics Professor even then. However, legislation was passed later in the nineteenth century which limited what could, and what could not, be a university. After that time, although it carried on awarding degrees, the degrees were Anderson University degrees validated by Glasgow University. I believe it also used to have a Medical Faculty, but this was fully absorbed into Glasgow University.

From the turn of the century, Strathclyde University was the only "non" University to be looked after by the University Grants Committee (which funds Universities) and it became well-known in the twentieth century as the Glasgow College of Science and Technology (established in 1912). It did not regard itself, therefore, as a new University when it was "re-made" as a university in 1965. At the time of writing, it is ranked 30th in the University research league tables. The Principal (or Vice Chancellor), when I was there was Sir Graham Hills, a man I greatly admired, liked and got on well with.

Once I was established at Strathclyde, I was approached by various Universities to become an External Examiner. I was appointed External

Examiner at City University (where I knew Norman Revell, with whom I had worked at IBM), at Loughborough (where I knew Ernest Edmonds) and at Heriot-Watt University (where I knew Mike Norman).

At Strathclyde there were two academics whom I had known for some years—the Head of Department, Prof Andrew Colin and Dr Douglas McGregor. Andrew was long established there and had pioneered the introduction of BASIC programming in Strathclyde very early on with the Commodore PET. He had also been a strong supporter of the microprocessor workshops at Liverpool, so I knew him well. When I applied for the Chair, Douglas McGregor competed with me for the post, and after the Vice-Chancellor had offered me the Chair at the interview, he asked me if I would have any objection to Douglas being given a personal Chair in addition to me being appointed to the main chair. I had great respect for Douglas and had no objections whatsoever, so he was made a professor as well.

On arriving at Strathclyde, I began to experience some of challenges that can occur in academic establishments. Unlike industry, different departments and different Universities do not always have the same goals. Of course, there are internal rivalries in all organisations, but at the end of the day, the key commercial goals will usually take precedence. Whilst Universities might have lofty overall goals such as a high level of teaching competence and excellence in research, there can be intense rivalry between universities, departments and even individuals in attaining these goals. Whilst at Liverpool I didn't experience this, mainly because I oversaw the Computer Department which served most of the departments of the University.

I immediately became caught up in a problem of cooperation between departments to achieve a common goal. That year, the Engineering and Physical Sciences Research Council (EPSRC) had been concerned that not enough graduates were going into Information Technology (IT). Computer Science departments were simply not large enough to provide all the IT specialists required and it was thought that graduates from other departments could be "converted" to IT specialists via a 1-year Master of Science (MSc) Programme (called an IT conversion course).

They therefore provided funding to encourage postgraduate students from other disciplines (Business, Science and Arts graduates) to move into IT. These studentships were fully funded, and students were awarded an MSc. in IT at the end of the course. I think they had their fees paid and received a grant of about

£2,000. One of the key conditions the EPSRC put on the award of the studentships (no doubt realising that there would be intense competition from different departments for the money) was that the course must be shared between Electrical Engineering, the Business School and Computer Science. How the money was divided was left up to each University.

The Head of the Department of Computer Science, Prof Colin, was very reluctant to go ahead, because of the politics involved in working with other departments. He had suffered a difficult experience earlier when collaborating with other departments in the University. However, I saw this as important since we could probably bid for about 30 studentships, so Andrew said, "You have a go at it"!

I took a different view of the politics involved. I pointed out that provided all departments were up-front about what contribution each department would make and this was agreed at the outset, things ought to work out. Andrew was very nervous about this but put me in charge of the negotiations. The three departments which would obviously be involved were Computer Science, Business Studies and Electrical Engineering. I had informal chats with the other two Heads of Departments, and it seemed to me that agreement was possible.

At the first meeting I insisted that the three departments had to agree a percentage share of the work for the course before any other decisions were made, and only then would we work out the detailed content of those sections. The three departments agreed to share out the work as follows—60% Computer Science, 25% Electrical Engineering and 15% Business which we all thought was acceptable. We then carved out the work on this basis, working within our own percentages, and there were no arguments whatsoever. Andrew was surprised and pleased and after that the cooperation between the three departments worked very well. We received 30 studentships per year, and the course commenced in October 1983. It ran very successfully for many years.

One of the consequences of the Microcomputer revolution, was that Clive Sinclair had already changed the nature of home computing with the introduction of the ZX81 and the ZX Spectrum personal computers, although these computers were too primitive for undergraduate computing. However, in 1984 Clive Sinclair announced the "QL" system for £399. QL stood for "Quantum Leap". This was a much higher-level computer than the Spectrum and was fully portable. It also had more advanced languages on it than BASIC (for example PASCAL) and it could be networked.

Strathclyde University had been a pioneer in using microcomputers for student teaching. (Prof Andrew Colin had already introduced PET computers) and the University currently had an 80-seater PET laboratory for teaching BASIC. At the same time Universities in the USA were proposing giving each student a personal computer, though costs over there were much less than in the UK. I therefore decided to propose a radical change in the provision of undergraduate computing at Strathclyde University.

On 20 February 1984, I wrote a paper entitled "A Quantum Leap in Undergraduate Computing" which I submitted to the University where I proposed to approach Clive Sinclair to suggest there might be a way of providing every undergraduate with a QL. I estimated that doing this could reduce the price to £200—£250, and that we might persuade Sinclair to give us some QLs.

I saw this facility initially providing students with access to lecture notes and teaching aids. In the early stages I thought that Computer Aided Learning might be too ambitious, though it could be provided later. A software house would also be created which could sell software to other Universities and Prof Colin was keen to create a company to do this. I contacted Sir Clive Sinclair and received a very positive reply.

I met the Sinclair team in October 1984, and they were very enthusiastic about my QL idea. They were keen to supply some QLs and we began negotiations. The Sinclair project progressed well during the rest of 1984, and I managed to persuade Sir Clive to supply us with 400 free QL systems, an astonishing number! Once I obtained these computers the rest was straightforward. I submitted and received £300,000 from the Computer Board to support the project.

We had some systems delivered and I took over one personally to see how well it performed. The only problem with the QL was that it had what were called, micro-drives rather than disks. These were not proper disk drives, but tiny magnetic tape drives. They normally worked OK but were occasionally unreliable and I worried that they might cause a problem with students. In November I gave a series of talks on the project and invited other Departments at Strathclyde to become involved. We appointed a coordinator who took the project forward from there. I was not involved after that, and many undergraduates became QL owners. Later, however, the QL was withdrawn from production as it failed to achieve the anticipated impact.

In retrospect, I think my proposals were probably too early. The disk drives caused more problems than I had envisaged but we were able to eventually replace them with proper disk drives. Today, of course, all undergraduates have laptops and a small lap-top or pad can be bought for as little as £200.

Human Computer Interaction (HCI) was still my main research interest although Mike Coombs was dragging me towards Artificial Intelligence (AI). The Japanese "Third Generation of Computers" initiative had caused the Government to panic over Artificial Intelligence and a committee—the Alvey Committee—had been set up by the Government to propose solutions. They recommended that huge amounts of money be invested into AI research. Interestingly, the Alvey Committee also recognised the importance of HCI and put additional money aside to fund research in that area, and they appointed an HCI Director—Chris Barrow.

Mike Norman (whom I had met previously in Liverpool) had, in the meantime, moved up to take a Senior Lectureship at Herriot Watt University in Edinburgh in 1980 and subsequently came to see me in Glasgow in autumn 1983, about a year after I had moved to Strathclyde. We discussed the possibility of obtaining grants in HCI from the Alvey Committee initiative and decided to go for a joint Centre (to be called the Scottish HCI Centre), based at both Strathclyde and Herriot-Watt Universities. The original name was the Scottish MMI Centre, but the term MMI (Man-Machine Interface) was replaced at that time by HCI (Human Computer Interface).

Mike and I got to work. We organised a one-day conference at Ross Priory by Loch Lomond on 9 December 1983, with the title "Specification and Design of Usable Systems". The objectives were to promote HCI in the Alvey programme, to identify expertise in both Universities as well as industrial and commercial interest.

In total, 29 people attended, including Chris Barrow, the Alvey HCI Director. Sir Clive Sinclair also attended, and there were delegates from ICL, Yard, BT, STL, Data Logic, Ferranti, the Scottish Development Agency, Britoil, SCICON, the Department of Trade and Industry, Southern General Hospital, CCTA and some civil servants. Chris Barrow thought the establishment of a Scottish HCI Centre was a great idea and he agreed to push it forward.

A first outcome was the approval, in 1984, of a project called Adaptive Intelligent Dialogue (The AID Project). This was with BT, STL Data Logic (all of whom had been at Ross Priory) together with Strathclyde (Me) and Heriot-

Watt (Mike). It was for a grant to support two research assistants, one at each University, a technician and a DEC VAX computer, valued at £338,593. This grant really established The Scottish HCI Centre and put us in a very good position to win further Alvey funding. The equipment was installed in August 1984. The AID project ran until summer 1987.

Late in 1984, Terry Mayes, a psychologist at Strathclyde, expressed interest in the work of the proposed HCI Centre. He came to work with us as Deputy Director (he was later appointed a Professor at Heriot-Watt). Terry was a great asset to the Centre.

The Alvey bid for the Centre involved major computing equipment, and four research projects totalling about £900,000, and in March 1986 it was announced we heard we had won the bids. A final memorandum of agreement was signed between Heriot-Watt and Strathclyde Universities on 2 July 1986, and the Scottish HCI Centre was born. The Strathclyde HCI Centre was set up in George House, adjacent to the Turing Institute. A similar Centre was set up at Heriot-Watt.

In the first quarter of 1986, just as we were starting up the Centre, Mike Norman was appointed to the Rank-Xerox Chair in Computer Science at Hull University. Possibly, Heriot-Watt should have seen this coming and offered Mike a Chair, but they did not. I have a copy of a letter from the Vice-Chancellor of Heriot-Watt University thanking Mike for his efforts and announcing my appointment as Head of the Centre. We were very concerned about the loss of Mike, but Heriot-Watt then announced that a lecturer at Heriot-Watt Computer Science Department would be appointed as Acting Director of the Centre (with myself as Chairman).

The lecturer who was appointed appeared at first to be a good choice, but things went downhill after that! I do not think I should name him so let us call him John Smith. There were discussions as to whether Mike should take some of the research grants with him to Hull, but the Alvey Directorate (and Heriot-Watt) were adamant that the grants should remain at the Centre. I personally would not have objected to, say one grant being transferred because Mike had contributed a lot in setting up the Centre, but Heriot-Watt were firmly of the view that this shouldn't happen. Mike finally took up the appointment at Hull on 1 October 1986.

Initially the Centre worked well but serious problems began to develop at the Heriot-Watt end. This is the trouble with universities, unlike IBM, where the main objective is to promote the organisation, many academic staff at universities seem to want to compete with each other! John Smith and I had never met before the Heriot-Watt announcement, and although he did not have a good research record, my initial meeting with John was a good one. Also, Chris Barrow (the HCI Director at Alvey) had recently left and his position had been taken up by Lawrence Clarke. Lawrence was quite a different person to Chris Barrow.

After this, things went from bad to worse. The first serious problem to occur was the appointment of the second programmer for the Centre. It was always envisaged that the Centre would have two support programmers, one at each site, to support HCI tool production. Gilbert Cockton who was appointed first at Heriot-Watt, was an excellent programmer and researcher and was an asset to the Centre. I then set about recruiting a programmer for Strathclyde.

At about this time, I met a programmer who worked for Hughes Aircraft in the USA. He was very interested in the Strathclyde position, and he seemed ideal for the post. He wanted to combine the post with reading for a PhD and I assured him that that this was possible. After a series of lengthy telephone calls (none of which involved Heriot-Watt) I offered him the post at Strathclyde, and the assumption was that he would start in January 1987.

In September 1986 he was in the UK and visited the Centre at Strathclyde, and I was away at the HCI Conference in York. I briefed my secretary to ask him to go to Edinburgh to finalise the contract. He was surprised because he had not realised that the Heriot-Watt part of the Unit actually existed. I asked him to go there since I thought it was appropriate that he met John Smith. In addition, Heriot-Watt, could set up a work permit for him. I heard no more, except that he would be starting in January 1987, and we organised a desk for him at Strathclyde.

In January 1987, I was in email contact with him, and I suddenly realised he was at Heriot-Watt University. I pointed out that he was expected to work at Strathclyde, but he responded by saying that John Smith had told him he could work in Edinburgh if he wanted and on whatever topic he chose! He therefore said he would not be moving. I do not blame him because he had understood in good faith that he could work from Edinburgh. The result was that Heriot-Watt had purloined the second programmer, and Strathclyde was now without a tool developer. I was really upset. I was the grant holder (Mr Smith was not) and had not even been consulted. The change in job specification was done without my approval, and as a result, Strathclyde could not now meet its Alvey tool development contract.

A few days later another even more upsetting event occurred. There was an IT86 Training Initiative from the Department of Trade and Industry (DTI) which offered money to provide equipment to support training. I found out that John Smith (again without any consultation) had been to London and had negotiated a £250,000 contract for Heriot-Watt with no mention of Strathclyde. He had posed the bid as a Centre activity but allocated the equipment only to Heriot-Watt. The bid proposed building on the training experience at Heriot-Watt, omitting to point out that the only industrial training that had been done to date was at Strathclyde! Finally, the bid had not been through the Centre Management Committee. The deadline for the bid was the day after I found out about it! There was no time to fight this move politically so I hastily wrote a completely parallel bid in the next 24 hours for £250,000 and submitted it to the DTI. Fortunately, the grants at both centres were approved.

This was outrageous behaviour and I complained bitterly to Heriot-Watt, but John Smith completely ignored my concerns. However, the Pro Vice-Chancellor at Heriot-Watt did see that we had been badly treated and tried to work out some form of compensation. From then on, I decided that there was no alternative but to separate the two components of the site, but to still have a cooperation mechanism which gave a common front to the outside world. However, at a Management Meeting on 19 June 1987, Lawrence Clarke of Alvey objected to this proposal saying that whilst the research could operate independently, it was important that the commercial activity was jointly done. Lawrence Clarke sent a 3-page letter suggesting a way forward.

On 2 July 1987, John Smith submitted a document to a meeting of the two Principals which contained some very odd remarks. For example, he criticised Strathclyde for a lack of identity and he criticised Heriot-Watt for integrating the Scottish HCI Centre into the Computer Science Department. In response, I pointed out to the Principals that Strathclyde no longer felt to be an equal partner, yet it had carried out most of the research, training and contact with industry. Finally, John suggested that he ought to be better rewarded for being Director of the Centre claiming that he was overloaded with lecturing responsibilities at Heriot-Watt. This was rebutted in a letter from the Head of Department at Heriot-Watt who pointed out that John only gave 41 lectures over the whole year and had no administrative duties.

In July 1987, things at the HCI Centre deteriorated further! We had a difficult meeting on 22 July, and John Smith appeared to have gone behind my back to see Lawrence Clarke and had not represented Strathclyde's position properly. However, the two of us met again on 24 July and hammered out a management proposal. This revised proposal was agreed with the SERC and accepted in a letter of 22 October. The contract was deemed to have started on 1 January 1985, and would terminate, as planned, on 31 December 1988. I hoped that this was at last the start of proper collaboration.

In September 1987, I produced a comprehensive plan to further develop the commercial side of the Centre by signing up more affiliates and further developing the commercial HCI library. On 16 October 1987, we held the first meeting of a newly constituted management committee for the Centre. It so happened that this was the day a hurricane struck the South of England, and many members were unable to attend.

The information given at this meeting was quite telling. For example, Strathclyde had sold 82 reports to industrialists in comparison with Heriot-Watt who had sold 15. Strathclyde now had the on-line HCI library (modelled on the Turing Institute Library). The reports had brought in an income of £1,238 and supported the creation of many contacts with industry both here and abroad. Strathclyde alone was running a set of seminars during the autumn term with several well-known speakers and had signed-up some industrial affiliates (these cost about £5,000 as a joining fee).

John Smith never accepted the absorption of the Heriot-Watt side of the Scottish HCI Centre into the Department of Computer Science at Heriot-Watt. On 6 November 1987, he sent an extraordinary letter to the Alvey Committee, which he copied to the two Principals but not to me. I still cannot understand why he sent such a letter. He opens with the sentence:

"I have hesitated to tell you about recent happenings at the Centre because I have been told not to contact you several times by Senior Staff at both Universities. But, as the incumbent Executive Director, I cannot allow the wilful destruction of the Centre. The enclosed dossier shows the amount of energy that has been wasted on what you have described as parochial matters; it was my last-ditch attempt to save the Centre." Finally, he delivers the most controversial sentence of all:

"Because the two Universities obviously do not share our commitment to the Scottish HCI Centre they must now forfeit their chance to be directly involved.

To be fair they should be given the opportunity to retire gracefully, whilst still allowing more collaboration. It will still be necessary to have a university provide a proper base for the academic research, and I would welcome suggestions on how to achieve this. So far, it has only been through me at Heriot-Watt."

The letter went straight to the Head of HCI at the Alvey Directorate, Lawrence Clarke. The first I heard of it was in a telephone call from the Technical Officer overseeing the HCI Centre for Alvey. He was taken aback by this letter and asked me what I thought of it, but I pointed out I hadn't even received it even though I was the Chairman of the Centre! He read it out and I had to sit down to absorb its contents! It was such a travesty I could scarcely believe it. Before ringing me, the Technical Officer had talked with Lawrence Clarke. Lawrence had said that Alvey wanted nothing to do with this!

Both Principals were apoplectic. The Principal of Strathclyde wrote to Lawrence Clarke on 10 November pointing out that my group has had to operate under great difficulty the source of which is now evident. He indicated that we remained strongly committed to the Alvey HCI Centre and had no intention of quitting the field.

Sir Graham wrote a letter to John Smith pointing out that the Scottish HCI Centre would survive this hiccup. Lawrence Clarke initially did not reply (I guess he was confused by the events). However, on 19 November he replied pointing out that he too "had regarded John Smith's letter with suspicion…" and that he was "glad that the Scottish HCI Centre was in good shape."

On 8 December, the Principal at Heriot-Watt University removed John Smith from his appointment as Executive Director of the Scottish HCI Centre and as HCI Group Leader/Director in the Dept of Computer Science. The Principal wrote a letter pointing out that "the two Universities were keen to see the HCI work succeed."

So, ended a most unfortunate incident which nearly wrecked all the good work of the Scottish HCI Centre. The Strathclyde side of the collaboration was regarded as a success despite the near initial disaster at Heriot-Watt. Many courses for industry had been held. Many reports had been written and widely disseminated, the two research grants that Strathclyde was responsible for had been completed successfully, and the HCI Library had been established.

Shortly after the above incidents, Prof Alistair Kilgour was appointed as a Prof of Computer Science at Heriot-Watt and took over the running of the Heriot-

Watt HCI Centre. I knew Alistair and we both got on well. From that moment, the two Universities worked well together and the Scottish HCI Centre was a great success. I felt sorry for the way Heriot-Watt had been treated by John Smith, and how this had caused difficulties for the Principal there because he had always been most supportive.

What happened to John Smith? I have no idea. I never saw him again. The Alvey contract for the Scottish HCI Centre finished as planned on 31 December 1988. During this time, the three research projects went well, and many research papers were produced.

An embarrassing incident happened to me about this time. The Chancellor of Strathclyde University was Lord Todd of Trumpington. I think he was born in Glasgow. He certainly spoke with a strong Scottish accent. I did not know him well at all. One night I was invited to a special dinner in the Principal's private dining room. It was probably after a degree ceremony where the Chancellor would have presided. There were about 12 people at this dinner with Sir Graham Hills, the Vice-Chancellor in the chair.

I was sat next to Lord Todd. I think Trumpington referred to the location of a village near Cambridge. We had a long conversation. I asked him what he did, and he replied that he had been involved in Chemistry. He then said he was fluent in a number of languages "Spanish, Italian, etc. and Chinese". I didn't take this seriously and asked him to speak some Chinese! He replied that he had been on the negotiating committee for the transfer of Hong Kong to China, so I thought I had better believe him. I think I then made some comment about Physics being more important than Chemistry, and he then made some comment about Physics. At the end of the dinner, the Vice-Chancellor rose to his feet. He said "We are all greatly honoured by the presence this evening of Lord Todd of Trumpington…" he paused and then said "Nobel Laureate for Chemistry 1955…" I could have died. I wanted to crawl away into some little-known space! I apologised to him for not knowing this, but he wasn't in the least bit bothered and said he had enjoyed talking to me. He was very kind!

In January 1989, the Centre had to submit a final report on the activities of the Centre over the four years of the Alvey Contract. Therefore, on 15 May, 1989, the Science and Engineering Research Council made a formal visit to Scotland to assess how successful the four-year Scottish HCI contract had been (the Alvey Committee had finished its work and had been disbanded). The meeting was held in Glasgow at the Strathclyde HCI Centre. The presenters for

Strathclyde were me, Dr Terry Mayes, and Peter Reid. From the Heriot-Watt side Prof Alistair Kilgour and Dr Patrick Holt presented.

The report was impressive (despite the difficulties in 1987). It listed 64 companies with whom the Centre had been in contact. It reported 51 research papers from Strathclyde and 46 from Heriot-Watt (notice how the Heriot-Watt contribution had improved after Alistair Kilgour took over). During the project Strathclyde had sold 336 research reports on various topics (to 24 Universities, 17 research establishments and 50 industrial concerns) and when the Centre finally settled down, Heriot-Watt also produced 89 reports (to 7 Universities, 5 research establishment and 7 to industry). All these reports were sold at prices between £5 and £30, and some were best sellers. For example, the £30 "Survey of Dialogue Systems and Literature on Dialogue Design" from Strathclyde sold 300 copies alone.

The two Centres listed 25 staff (Strathclyde) and 24 staff (Heriot-Watt). Many demonstrations of HCI tools were given. The visiting panel were impressed by the results of the four years' work and gave us an excellent final report. I was relieved that, in the end, the Centre had been a success, but I was heavily bruised by these events and vowed never to enter contracts with other Universities again!

On a totally different matter, whilst I was at Strathclyde, I also become involved with the Ministry of Defence. In 1982, Professor Keith Bowden (a Computer Scientist at Essex University) died in a car accident when driving home from London. Police said it may have been drink related, but his wife claimed the car had been tampered with. Prof Bowden had been working on the USA "Star Wars" project and was Chair of a UK Defence Committee. There had also been other mysterious deaths of other Star Wars scientists around the time, and there were suspicions that Prof Bowdon had been killed perhaps by Soviet agents. The issue was never resolved but, because of Keith Bowden's death I was contacted by Prof Bob Churchhouse, of Cardiff (who was also Chair of the Computer Board) to see if I would replace Keith on a UK Defence Committee. I signed the Official Secrets Act in October and served on the Committee for the next ten years eventually becoming Chair of the Committee.

For obvious security reasons I cannot report what I did with this committee. I mainly worked with the Royal Navy and always found naval people very interesting to talk with. One interesting event which I can tell you about happened in Sept 1983 we visited the Royal Engineers at Chatham. Apart from the work we did there, we were shown their museum. I hadn't realised that it was the Royal Engineers who defended Rourke's Drift (made famous by the film Zulu—eleven Victoria Crosses (VCs) were awarded at Rourke's Drift). In the evening we attended a regimental dinner and afterwards saw an extensive model of Rourke's Drift in the basement.

Chapter 18
The Turing Institute for Artificial Intelligence

Just before I left Liverpool, I became involved with Artificial Intelligence (AI) Research. Artificial Intelligence (AI) had suffered a rather chequered career after Alan Turing's initial paper. An interesting test of the Turing Test happened in 1964 when Joseph Weisenbaum (German American computer scientist) created a program called ELIZA which mimicked human conversation. It had no intelligence in it whatsoever, but simply reflected back what the user had asked in a different form. So, if a user asked, "How is your father" It might reply "Are you worried about your father?" If it was really stuck for a reply, it would say," "Please go on." This mimics the human trick of replying "uh huh" when someone is telling you something. It encourages them to give you more information. The program often survived the Turing Test for several minutes though it tended to fail if there were any hidden messages in the questions of the user. Also, it didn't actually do anything!

Later problems with AI began with the publication of a report by Sir James Lighthill in 1973 entitled "Artificial Intelligence: A General Survey". It was compiled for the British Science Research Council and stated, "in no part of the field have discoveries made so far produced the major impact that was then promised".

As a result, the British government ended support for AI research in all but three universities—Edinburgh, Sussex and Essex. The report was highly critical of basic research in foundational areas such as robotics and language processing. The report concluded that whilst AI techniques might work within the scope of small problem domains, they would not scale up well to solve more realistic problems. There was a major debate between Sir John Lighthill and Prof Donald

Michie in 1973, who challenged the report's findings, but AI research was severely curtailed for number of years.

Towards the end of the seventies the research interest picked up again mainly led by work in the United States. Lighthill was certainly correct when he suggested that AI might work best in small problem domains because of the difficulties in trying to solve large scale problems and it was realised that suitable domains could be areas of expertise which involve a high level of expertise in a narrow area.

Towards the end of 1981, shortly before I left Liverpool, Mike Coombs came to see me and said that he had come across some important developments in AI in the States which we ought to get involved in. He said they were called "Expert Systems" (or Intelligent Knowledge-Based Systems IKBS). I had never heard about them, but we began to read about the subject and at Christmas I offered a bottle of wine for the first of us (including myself) to write a small expert system. I wrote a small system to sort numbers in ascending order and won the bottle of wine (Mike was not really a programmer!). We then decided to go to the United States to find out what was happening in the area and approached the National Computing Centre (NCC) for funding. They agreed to fund us provided we wrote a book of our experiences on our return.

We went over to the United States to meet with the main researchers. Most developments were in the medical field. We met Edward Shortliffe (who developed an Expert System called MYCIN to assist clinicians in the diagnosis and treatment of infectious diseases), Richard Duda (who developed PROSPECTOR to assist field geologists), and we met other developers (Myers) who had designed CASNET (diagnosing Glaucoma) and INTERNIST (which was concerned with the Diagnosis and Manifestations of a disease).

As soon as we arrived back from the USA, I took up my new position at Strathclyde. Mike and I wrote up our visits to the Expert Systems people in the USA and submitted it, as a book, to the National Computer Centre. We were not real experts, but we thought the work was important, and in any event, we had promised to write up the results of our visit in the form of a textbook. They approved the book and in 1983 our book (right) *"Expert Systems: Concepts and Examples"* was published by Wiley. It turned out to be very popular indeed for undergraduate

courses in Artificial Intelligence (AI) and sold over 35,000 copies. The book was later translated into French, Spanish and Japanese (unfortunately authors get very limited royalties for translations!) and later, when I went to Moscow, I discovered that it was a very popular University book in Russia though we never received any royalties!

Expert systems were programmed very differently from conventional programs. They used what are called "Production Systems" which contained sets of Production Rules of the form, *IF condition THEN action* which act on a knowledge base. The production system then searches for the most appropriate rule to apply in particular situations, as the knowledge search progresses. The other advantage of Production Rules was that they are expressed in the language of the expert rather than in an abstract (computer programming) language so experts can understand them. So, a set of rules might look like:

IF (Current=0) and (Plug is in) THEN Switch On.

IF (Switch is On) and (Current = 0) THEN Check Fuse

The control structure searches through the current data values (Current, Switch, Plug) and fires the appropriate rule. It will then search again through the rules and trigger other rules. Often expert knowledge can be expressed in sets of such rules, and experts like it because the language is familiar to them. The production approach can work quite well with well-defined expertise, but it does not work so well with what we might call everyday knowledge or common sense.

Interest in Artificial Intelligence (AI) grew rapidly in the 1980's partly driven by the work done in Japan. As a result, the British Government set up the Alvey Committee to support AI research in the UK, so when I went to Strathclyde the world had become very positive about the possibilities of AI.

One successful development in AI at the time which is used a lot today was the introduction of Object-Oriented Programming Languages (OOPS). Knowledge is incorporated into objects, and they can inherit knowledge from higher level objects. In standard programming the variables are separately defined and are then manipulated in the program to give the desired result. In O-O programming, computer programs are designed by making them out of objects that interact with one another. The idea really came out of Artificial Intelligence programming since objects have knowledge of themselves. In Object-Oriented programming (OOP) the object is the main data structure. Objects contain data, in the form of fields (or attributes) and they also contain procedures (or methods).

An object's procedures can access and often modify the data fields of the object with which they are associated (objects have a notion of "this" or "self").

The first Object Oriented Language was SIMULA67, but the term Object Oriented Programming was first introduced by Alan Kay in his Smalltalk language, developed at Xerox PARC) in the 1970s. The term *object-oriented programming* represented the pervasive use of objects and messages as the basis for computation. The Computer Magazine BYTE introduced Smalltalk to the world in August 1981.

Objects sometimes correspond to things found in the real world. For example, a graphics program may have objects such as "circle", "line", or "rectangle". An on-line reservation system might have "seats", "planes" or "airports". Objects can be abstract as well as real entities (i.e., an open file).

Languages that support classes usually supported "inheritance". This allows classes to be arranged in a hierarchy that represents "is-a-type-of" relationships. For example, class Employee might inherit from class Person. All of the data and methods available to the parent class also appear in the child class with the same names. The program is therefore storing not just the variables but the knowledge and relationships between the objects.

For example, class Person might define variables "Forename" and "Last name" with method "make full name". These will also be available in class Employee, which might add the variables "position" and "salary". This technique allows easy re-use of the same procedures and data definitions, in addition to potentially mirroring real-world relationships in an intuitive way.

I had been at Strathclyde about a year when the Principal, Sir Graham Hills, announced that Prof Donald Michie and his team were moving to Strathclyde from Edinburgh University and were setting up the Turing Institute for Artificial Intelligence (AI) in memory of Alan Turing in Glasgow. The Institute was being established in George House, which is in George Square, right next to the University, and would be independent but closely connected with the University (for example, it would be financially independent of the University). The actual move was planned to take place the following year in 1984. Sir Graham said that he wanted to make sure that this Institute was properly run and asked me to join the Turing Board as the Strathclyde representative to "keep an eye on it!" I readily agreed because I had heard quite a lot about Prof Michie and I knew that he had worked with Alan Turing at Bletchley Park. I was therefore delighted to join the Turing Institute Board in the summer of 1983.

The Chair of the Board was Lord Balfour of Burleigh, who was also Deputy Governor of the Bank of Scotland. He was a person with whom I got on very well, and we had a happy association for the next few years. I was very interested in becoming involved, as it was an Artificial Intelligence Research Institute and Mike Coombs and I had become very interested in the area. At the same time, I had just been made Head of Department of Computer Science which frankly I didn't particularly enjoy doing.

Prof Michie visited me in December 1983 and in January 1984 I met two of the main researchers who were moving to the new Institute from Edinburgh, Dr Peter Mowforth and Dr Tim Niblett. I also met Jim Richmond, who had been appointed as the General Manager of the Turing Institute. We had a number of Board meetings, and the Institute was set up in George House early in 1984.

In the summer of 1984, Prof Michie reported that he had a heart condition and couldn't continue as Director of the Institute, and I was asked by the Principal to take over as the Executive Director of the Institute. I never found out whether Prof Michie's illness was real or not—he seemed remarkably well afterwards! Also, I had never been happy as Head of Department of Computer Science, so I accepted this new post and resigned as Head of Department (which I was very glad to do) but I still carried on as a full-time Professor at Strathclyde. Strathclyde agreed to move all my HCI research staff into George House so that I could run both the Turing Institute and the Scottish HCI Research Centre under one roof.

When I began as Executive Director, the Institute even provided me with a car! I started my five years as Executive Director just as Mr Gordon Pattie, Minister for Industry, officially opened the Turing Institute in George Square on 12 December 1984. In the autumn I quickly found that the Institute had nearly run out of money (was this the reason for Donald's heart condition!). I met with Jim Richmond, the Office Manager, to sort out the problems. Jim was a great manager, and I relied on him heavily during the next five years. The Institute was a company limited by guarantee, which meant it could not go into the red at the end of the year and in the year that I took over as Director we were precariously near to going under.

The Institute obtained its money in three ways. Firstly, companies could join as affiliates of the Institute for £15,000 per year. Affiliates could send people to the Institute for in-depth training (called Journeymen) and they had on-line access to the Institute's first-class Artificial Intelligence on-line library. The library really was excellent and was the brainchild of Donald Michie. All the most up-to-date papers and books in AI were there and could be accessed by affiliates on-line. Secondly, members of the Institute gave commercial courses. Thirdly, the Institute submitted bids for Research Grants to the SERC, to ESPRIT (European funding) and to other bodies.

In 1984, ICS (an American Teaching Company) approached me as asked me to lecture on Artificial Intelligence (AI). I didn't really regard myself as a real expert, but I had worked with Mike Coombs on AI related work for about 3 years. I gave an initial course in June 1984 and in September 1984, they sent me to Palo Alto in California to see the four-day course being given by the originator, George Luger. Then, in 1985 I went around the UK giving two and three-day Expert Systems courses which brought in good money for the Institute and advertised its existence. Early on, although I knew a lot about Expert Systems I did not class myself as a true expert, but by repeatedly giving these courses I eventually did become a real expert! I also gave courses on the logic language PROLOG.

I gave many courses in 1986 mainly to earn money for the Institute. I gave three-day courses to Smiths Industries, British Gas, British Telecom and IBM. I gave a four-day Expert Systems course in Heidelberg and lectured at several international conferences. I also gave many courses in Glasgow on Expert Systems at the Institute. Whilst I was paid for these courses, I donated most of the money to the Institute to assist in solving their financial difficulties. In parallel we made various applications for grants, and I made contact with YARD and Britoil in Glasgow, and they joined the affiliates scheme. It was clear we were beginning to turn the Institute around financially.

In February 1996, an incident occurred live on BBC television in which I illustrated how human beings acquired knowledge using what is known as rule induction and it had an amusing end! The BBC approached me to take part in a live television programme called MicroLive. This programme went out live every Friday night and discussed computer technology topics. It was quite a popular programme and included Lesley Judd, a presenter from the Children's programme "Blue Peter". They wanted me to appear since the topic on this particular night was Artificial Intelligence. The programme was broadcast on 21 February 1986, and I had to go down for the day to rehearse before the live broadcast at 6pm. In those days many television programmes were actually live. We rehearsed all day. My main appearance was in a live interview with Iain

McNaught Davies (known as "Mac") discussing the Technique of Induction and using the Space Shuttle as an example (in which the Institute had been involved).

Induction is a common way in which human beings find out about the world. We slowly build up a set of rules governing our behaviour from real-world experiences, and we modify the rules as more information is gathered.

During the practices for the interview, Mac asked me to describe how people gained knowledge about the world (i.e., Induction). I used a real example of what happened to me when I first went to Cairo in January 1980 working on Databases. I was driven into the city by taxi from the airport. We approached a set of traffic lights, which were on red, and the driver drove straight through. I asked him why he had done this, and he replied "We don't take any notice of traffic lights in Cairo. I have a copy of the Koran in the back of the taxi. This protects me." I hoped he was a man of faith! This allowed me to induce the first rule "Traffic lights are ignored in Cairo."

At the next set of traffic lights, they were on red, and he stopped! I asked him why and he replied, "There is a policeman in the centre of the junction, and I will get fined." I was therefore able to refine my induced rule to "Traffic lights are ignored in Cairo unless there is a policeman at the junction."

Later we arrived at a junction with the lights on red. There was a policeman at the centre of the junction, and we drove straight through! I asked again why he had done this. He replied, "The policeman did not have his notebook out!" and explained that the policeman would write down the numbers of the cars which ignored the lights, and they would later receive a fine. However, if the policeman did not have his notebook out, he could not record the car number! The induced rule was therefore further modified to be "Traffic lights are ignored in Cairo unless there is a policeman at the junction, and he has his notebook out."

It was quite a nice example of progressive induction which human beings do all the time. So, by induction the rule is gradually built and modified through experience. Is it the correct rule?—you can never be sure, but it does cope with a situation where knowledge changes over time. The producer loved the story and said we must include it in the programme.

At the end of the first rehearsal, Lesley Judd held up a textbook on Artificial Intelligence by Prof Michie and suggested that viewers might wish to buy this if they wanted to know more. I objected to this and said, "I have written a book on Artificial Intelligence—why don't you use my book instead?" They agreed to go out and buy a copy. In the meantime, because they didn't know the title, during later rehearsals, the teleprompter which guided Lesley was altered to read "If you want to know more about Artificial Intelligence, you may wish to buy a copy of Prof Alty's book entitled "A Guide to the Red-Light District of Cairo"!

As the rehearsal went on Lesley said to the Producer, if you don't alter that to the correct title, I will say it when we are live. In the end they finally corrected the teleprompter, and the programme went really well, and Lesley managed to say the correct title "Expert Systems: Concepts and Examples".

However, the use of the original bogus title had later unintended consequences! After the programme, a female viewer wrote in saying that she had failed to catch the title of my book. A secretary had gone to the file and lifted out the old teleprompter sheet and had written back to the viewer with the title "A Guide to the Red-Light District of Cairo". The BBC had to apologise profusely both to the woman and myself, though I found it rather amusing!

So, the rules developed through induction give us an everyday understanding of the world as distinct from "logical" or "rigorous" theories. Of course, they are not necessarily correct but are improved through experience.

I have previously mentioned the Alvey Programme. We were fortunate at this time that the Alvey Programme had just been announced which initiated a major change in the way computing research was organised in the UK as a whole. The Japanese "Third Generation of Computers" initiative (which specifically stressed the importance of Artificial Intelligence) was mentioned in the previous chapter and had caused the Government to react by forming the Alvey Committee to propose solutions. They recommended that huge amounts of money be invested into AI research. The funding was substantial—about £320 million at 1982 prices. The programme ran for four years between 1983 and 1987. In the previous chapter I have already mentioned that the Scottish HCI Centre was set up by the Alvey Directorate and the Turing Institute benefitted as well. We obtained a number of research projects during the next few years, one of which was a project called PIMMS project which later caused me major problems as you will see.

In 1986, The Sunday Times rang me and said they were going to do an article on Artificial Intelligence in the colour supplement of their newspaper. Their idea was to show four of the well-known people in Artificial Intelligence in a natural setting. They asked me for an appropriate setting, and I suggested skiing! In June a photographer (Margaret Fear) came up to Glasgow and we went to the Cairngorms where, incredibly there was still plenty of snow. It was quite windy, and as we went up the first chair lift, we passed some sheep with their backs to the wind. The wind was parting their fleeces along their backs. Margaret noticed this and said how interesting it was, and she took many photographs of me in my blue ski suit. When the edition of the Sunday Times appeared, there was a full-page photo of me and incredibly, she had managed to get my hair parted by the wind, just like the sheep! In the photo, it is not obvious that I am standing on snow, but I was. The photo appeared in the Sunday Times Magazine of 20 July 1986 (right).

The other four people in the article were Bob Kowalsky of Imperial College (pictured in the bluebells), Karen Spark-Jones of Cambridge (on a bicycle) and Donald Michie. Donald did not have any sporting interests, so they pictured him with his robots! The article was not very interesting, but it did give the Institute some good publicity.

On 30 October 1986, we instituted the Turing Memorial Lecture, to be held annually at Glasgow, given by someone who had made an outstanding contribution to Artificial Intelligence. The first year the lecture was given by Alan Robinson who had made major contributions to Logic Programming. He gave a wonderful lecture to a packed audience. I remember him coming to me before the lecture embarrassed because he wasn't using transparencies (a sheet of celluloid on which you could write with coloured pens and then get them projected onto a large screen – everyone used them at that time!). I said it didn't matter, but he dutifully made a single slide for use in the lecture. At the beginning of the lecture, he pointed out that he didn't normally use slides, but he had actually prepared one! He stood before the podium, leaning on it with his back and delivered a lecture which held everyone absolutely spellbound. Just before the end of the lecture he looked at his prepared slide and said, "I guess we don't need this!" and threw it away! It was a great example of how you don't need all the fancy lecturing aids that people generally use today.

We didn't have a 1987 lecture as far as I remember, but Keith Clarke gave the 1988 Turing Memorial Lecture and the next two annual lectures were given by two other famous AI researchers—John McCarthy gave the 1989 Turing Lecture, and Herbert Simon gave the 1990 Turing Lecture.

In January 1987 a crisis occurred in one of the Institute's ESPRIT projects just as we were becoming financially sound. The project was called PIMMS (I can't remember what it stood for) and I had not been involved in setting it up. The Institute was accused by the other partners of not contributing fully and we were threatened with expulsion. It was an important project financially and I flew to Grenoble on 21 January 1988, to a project meeting to save it. On the second day they had a project meeting, and it was probably one of the most uncomfortable meetings I have ever attended. I argued throughout the afternoon, and I eventually convinced them that we should stay in the project (I was not totally convinced of the value of the other contributions either!). On the last day we skied at a very small resort above Grenoble, and I came back satisfied that I had saved the project.

However, Donald, for some reason, was unhappy. He suggested that it was a failed project and that we should withdraw immediately. I pointed out that it would seriously endanger the finances of the Institute and refused to pull out. Donald then sent me a letter saying that I was compromising the integrity of the Institute and if I did not pull out, he would write to the Chair and the rest of the Board Meeting complaining of my attitude. I therefore called his bluff. I immediately copied the letter to all members of the Board and put the item on the agenda for the next Board Meeting. Donald had to back down at the Board and expressed full confidence in me. I don't know why he did it, but he did tend to occasionally let his heart rule his head!

The combination of HCI in the Scottish HCI Centre and the Turing Institute proved to be most productive. I was successful in obtaining a large Research Grant (£350,000) from ESPRIT which combined both HCI and AI techniques. This work is dealt with in Chapter 20.

In November 1988 I decided to try to get funding from the European Space Agency. They had put out a proposal for bids to solve the problem of diagnosing faults on a spacecraft. They had the idea that an on-board expert system might work. At the Institute we had recently developed a technique for quantitative modelling of processes and, by progressively introducing faults we had induced the rules for dealing with them. This seemed to me to be an excellent approach for the Space Agency problem, and I flew to Nordvijk in Holland to present our approach. They received it enthusiastically and we won the contract. —£80,000. Danny Pierce at the Institute took over the running of the contract and it went very well indeed.

On 18 December 1988, I flew to Tampere to give a four-day course on Expert Systems for Nokia. It was very cold (-32 degrees C). The course went very well and one afternoon they took me out to a lake near Tampere where there was a Sauna on its banks. When I arrived there, I couldn't believe it. The Lake was frozen solid, and we crossed the ice in a snow mobile. I distinctly remember the ice cracking and my companions were debating whether we should get off the lake or not! However, we eventually arrived near the Sauna. "What do we do for water? It's completely frozen" I asked. "We go into the lake" They replied. "But there is 8 inches of ice on it" I said. "You'll see" they said!

When we arrived, I saw that there was a round hole in the ice (about 15 feet in diameter and the ice was about 18 inches deep). From the Sauna you walked down some duck boards and jumped in. There was a metal ladder so you could get out of the water. A continuous steam of bubbles was breaking the surface. "What are the bubbles for" I asked. "They are to stop the water icing over. The air temperature is -32 degrees! Without the bubbles, it would freeze immediately."

They explained that we would go into the sauna, get hot and sweaty and after that we would go down the duckboards from the sauna and jump in the lake. "But it is -32 degrees," I said, "It will be too cold."

"No," they replied. "The water is at 0 degrees C so jumping into the water from an air temperature of -32 C is like entering a hot bath." I believed them and having gotten very hot, I jumped in the lake. It was absolutely freezing!

I quickly grabbed the steel rail to pull myself up and my hands began to freeze on the rail! "Don't keep hold or you will freeze to the rail," they said. "Then we will come back for you in the summer!" I reached the duckboards and could feel my feet trying to stick to the surface. A generous cup of vodka revived me. Afterwards, I said to the Finnish friend who had brought me to this Sauna, "I have never experienced that before."

"Neither have I!" he replied!

We had lots of interesting visitors to the Institute. Prof Ivan Bratko came over from the University of Ljubljana and had written a number of books on AI. I have already mentioned John McCarthy (Inventor of the LISP programming language). Steve Muggleton also came up from Cambridge to visit regularly. He went to Imperial College. Another well-known person was Arthur Van Hoff, originally from Holland. Arthur was a brilliant programmer and really made a difference at the Institute. Later he moved to Silicon Valley, and I think he became a millionaire!

Donald Michie did a lot of work in Robotics. I remember him balancing poles and it was very effective. Robotics is an interesting subject, but I have never understood why AI people persist in making robots like human beings. Why give a robot two eyes when it can have seven! Why insist on two legs when a robot can have six!

Early in 1990 I was anxious to get back (full-time) into HCI research. I also felt that the Turing Institute required a full-time Executive Director. BT had decided to fund a Professorship at Strathclyde associated with the Institute. Since I had wanted for some time to give up the Executive Director Role, the Institute looked for a replacement for me. We interviewed several people and we eventually appointed Jack Donald. I was then appointed BT Research Professor and Jack took up his full-time appointment on 12 March 1990.

In fact, my desire to give up the Executive Directorship was more specific. I was increasingly keen to move back to England to be nearer the family, and to move full-time into HCI research. Prof Ernest Edmonds of Loughborough University, who was well-known for HCI research had already approached me in January 1990 indicating that there might be a Professorship at Loughborough University. I was asked to a special interview (I was the only candidate) and they offered me the Chair of Computer Science at Loughborough which I accepted. I think Jack Donald was quite upset because he had been relying on me as Research Director. However, the Computer Science Department at Loughborough was coincidental with my main research interests and there were many well-known people there—Ernest Edmonds, Steve Scrivener, and Brian Shackel (who ran HUSAT – The Human Sciences and Advanced Technology Centre at Loughborough).

Ernest Edmonds was delighted. I had known Ernest since he first came to the HCI workshop which Mike Coombs and I had organised in 1979 at Liverpool. I was also pleased because I was joining a university which was very strong on

HCI research. I informed the Principal of the University and The Institute. Both Donald Michie and Sir Graham Hills were quite taken aback by my impending departure, but Sir Graham realised that I wanted to move on. I also knew that Sir Graham was retiring as Principal and Vice-Chancellor at the end of the following year. I had always found Sir Graham really helpful and very dynamic, so it did worry me as to who would replace him and what would my relationship with them be?

In September 1990 shortly before I left the Turing Institute, Peter Mowforth organised the *1st International Robot Olympic*s at Glasgow hosted by the Turing Institute. The idea was publicity for the Institute and Robotics in General. The response was amazing. We had entries from Russia, Germany and many other countries. TV stations across Europe were also interested. We had events such as "The Five Yard Dash" and "The Wall Climb". Quite a lot of people attended the sessions, and it was overall a great success.

I finally finished full-time at the Turing Institute at the end of September 1990. On 27 September, I held a dinner for senior member of the Institute. Lord Balfour came, Sir Graham and Lady Hills, Peter Mowforth, Tim Niblett, Prof Michie, Terry Mayes and their wives. It was a very pleasant affair. So, at that dinner I said farewell to the members of the Institute, with whom I had really enjoyed working, and wished them well.

Chapter 19
Presentations: If Things Can Go Wrong, They Will!

Computer professionals are regularly asked to give papers to international gatherings, conferences, or to courses or meetings. If the professional is also an academic, they will also be lecturing under-graduates, post-graduates and staff of universities across the world. Giving presentations in foreign parts can be tricky, but even at home there are disasters which can happen. This chapter describes some of the problems that have occurred to me whilst presenting and travelling.

In the early days, presentations were by Flip Chart or by 35mm slideshow. Later foil projectors were used. With a Foil Projector, the Slides are actually transparent A4 sheets on which the speaker can write the information to be presented with coloured pens, or the foils could be produced and printed by a computer. To display the foil, the speaker places the slides on the foil projector which projects them onto a large screen. More recently slides are usually created on the computer and projected directly on to the screen.

Such presenting might seem quite straightforward, but it can be quite a hazardous business at times! Lots can go wrong and often will go wrong. Here are some of the nightmare scenarios that have happened in my career.

Not allowing enough time before the presentation can cause all sorts of problems. In 1979, shortly after I wrote the "Alty Report" on Microtechnology, I was asked by IBM to give a 1-hour presentation to about 70 Senior Managers on *Micros and their Implications*. The location was prestigious—the Manchester United Board Room. As you can imagine it was a very upmarket venue and it had 2 inches of thick carpet on the floor. I had already given many presentations on Microtechnology during that year, so I had about 100 typed projector foils from which I could build a presentation. Usually, I arrived in plenty of time and chose a set of slides from the set of 100 to tailor the presentation to the audience. Typically, I might choose 20 or 25 slides. The last presentation I had given would be on the top of the set.

I did not allow enough time to drive from Liverpool to Manchester and I arrived a little late. I walked into the room next to the Board Room, just in time to hear the Chairman announcing me in the adjoining room. There was no time to select the slides to be used in the presentation, so I decided to walk in with all the slides because the last presentation I had done would be on top of the set of slides. I thought I could use that and improvise a little. I had to enter at the back of the room with the delegates having their backs to me. There were about 8 rows of delegates with about 15 on each row.

As I rushed into the room, I failed to see that there was a piece of wood across the bottom of the door frame. My feet hit the wood and I came flying horizontally into the room headfirst. I made a perfect landing on the thick carpet and the foils came out of my hand and spread on the floor all the way up to the front. The delegates "helped" me by picking up the foils nearest to them and handing them to a collector. He gave them to back to me. What this effectively did was to randomly order the slides! I now had a completely random set of slides in my hand, but I had immediately to begin the presentation, so I picked myself up, strode confidently to the podium and I just put the top slide on the projector, hoping it might be relevant! This was how I carried on. Each succeeding side was a slight surprise, but I managed to make a coherent story about it. In fact, there were only a few slides which I passed over, and of course it was lucky that I was completely familiar with the content. Ironically, it was one of the best presentations that I gave that year because I had to think on my feet!

Misunderstanding what the meeting is about is another hazard. I was invited to Northwick Park Hospital, near London (again to talk on Microtechnology). The plan was to arrive for lunch and then (I thought) I would meet with the research team in a small committee room and discuss microtechnology. We finished lunch and made our way across the room to the next corridor. There a door was opened which I thought was entering a small office to meet the research team, but to my consternation it was a Lecture Theatre for about 250 people, and it was full—indeed people were sitting on the stairs. I suddenly realised I was giving a 1-hour lecture to about 300 people! Fortunately, I had my 100 foils with me, and the last presentation was still on the top, so I gave that (with a few variations!).

In 1985 I was invited to give a 3-hour tutorial on Expert Systems at West Virginia University. The seminar consisted of two 1 ½ hour sessions. I checked the lecture theatre before the presentation, and all seemed fine. It was a typical lecture theatre with tiered seats and at the front was a long desk behind which the presenter could stand to deliver the presentation. On the desk was the foil projector and a water jug with glasses. I do not like presenting behind a desk, so I moved the projector to the side of the desk, and I stood on the platform to present. I had a microphone attached to me and the presentation was videoed, and copies were later available for sale to participants. The lecture theatre was packed, and people were sitting on chairs very close to me. The presentation began well. About ½ an hour in I put a new slide on the foil projector and to make a point banged the top of the desk. What I hadn't realised was that the desk was not a permanent feature of the lecture theatre. It was a long table with very thin legs, and they had covered it with some coloured material. As I banged the desk the thin leg nearest me came off the platform. The desk, the foil projector, my slide, AND I, catapulted into the audience. The whole event was captured on video (including some of my comments!). Luckily the water jug and glasses were metal, so they did not break, but the water went onto the floor and over some of my slides. A few of the slides I had made the previous evening using coloured pens with soluble ink and I could see some of my slides dissolving in front of me. Members of the audience hauled me back onto the platform and luckily the destroyed slides were not important. I was told afterwards that the video was the best-selling one at the conference! When the second session started some of the members of the audience moved their chairs back as I began! I received a very good rating from the audience for the presentation (though was it really for the presentation!).

The rating given for presentations are often variable depending on the audience nationality. Americans always gave more positive ratings than Europeans. They were often happy to give the presenter a top rating. Europeans rarely did so, though they would always give a reasonable rating.

Even giving a vote of thanks can cause problems! In November 1987 I went with Prof Michie to Pittsburgh to visit Westinghouse, one of our Turing Institute Affiliates. I had a bad cold and didn't feel particularly well. I don't recall much about the two-day visit, but I remember flying back overnight and landing at Heathrow in the morning and catching the shuttle flight up to Glasgow. I called in the Institute feeling very tired and had planned to go home. Prof Michie had invited the British Chess Grand Master (Nigel Short) to give a seminar in the Institute at 2pm. He asked me to give a vote of thanks at the end of the seminar.

I was incredibly tired and sat at the back of the seminar (which was crowded). I told Tim Niblett (one of the illustrious researchers at the Institute) that I was very tired, and I would be shutting my eyes. Could he please nudge me when Nigel Short finished? I then fell fast asleep. Tim nudged me when question time started. I awoke and saw Prof Michie looking at me curiously and Nigel Short was silent, so I assumed that the Vote of Thanks time had arrived. I jumped to my feet and started "I would like to thank Nigel Short for this excellent…." and Nigel interrupted me "I haven't answered the question yet!" and I realised that I had spoken too soon! I sat down, now wide awake, and finally at the end had to get up to start the Vote of Thanks again. The whole room roared with laughter because they realised what had happened. The next evening, I had dinner with Nigel and Donald Michie and we had another good laugh at the incident.

Also, strange things can happen to you when you travel abroad, and a classic problem is an English phrase which is obvious to you but means something different to your foreign hosts. In the late 1980s I was asked to give an "After Dinner Talk" to the Danish Academy of Sciences who were having a Conference in Copenhagen. I arrived at about 5pm and was told that dinner was at 6pm. I thought that a seemed a bit early but didn't question it. I had prepared a typical after dinner talk of about 20 minutes with some amusing anecdotes in it.

When we sat down to dinner, it was informal. All the tables were for 4 people and there was no "high" table. I began to worry about where I was going to speak so during the desert course, I asked my host where I should give my talk. He looked surprised! "You are not speaking here. You will be lecturing in the Lecture theatre after dinner. We are all looking forward to your views on Artificial Intelligence!" I then learned that in Denmark an "after dinner talk" meant a formal talk after dinner in a lecture theatre! This time I did not have any slides with me. I said, "get me three blank foils and some coloured pens" During coffee I managed to plan out a one-hour presentation and then immediately had to enter the lecture theatre. In the end the talk went quite well since, being Executive Director of the Turing Institute I did know the subject. But it was a close call, particularly to such a distinguished audience.

I went to give a paper at a conference near Moscow in 1980. The conference was quite enjoyable, but the food was poor and one night, a Russian Professor came up to me and asked if I would like a trip into the Russian countryside. He said his parents were typical peasants and we would see how they lived. We might even get a good meal! He commandeered a car and we set off and drove into the Russian night. It was snowing lightly and very dark. We then picked up the parents and there were now seven of us packed into in the car. We drove for about one hour and then suddenly drove off the road into a field. An argument started in the car in Russian. My colleagues informed me that we were probably driving over a lake and the driver was not sure if the ice would take us! We survived and arrived at a tiny house.

They gave us hot mugs of fat to drink and then we had a meal in the house with plenty of vodka, standing and clicking our heels and downing the drinks in one go. After the meal they took us next door and we were in a room with grass on the floor and a rabbit warren in the centre. It turned out that the walls of the house (which were wooden planks) went deep into the earth and trapped the rabbits, so they had a captive warren of rabbits, and we had just eaten one! The chimney was also interesting. It rose vertically, and then went horizontal for about eight feet and then rose again vertically again. In the U-shaped recess was the bed, so the occupants were nice and warm during the cold night.

Problems can occur when there are people in the audience who know more than the presenter. I gave a course late in 1988 at the Turing Institute on Expert System Shells (these are programs which make it more straightforward to create an expert system). I gave many examples of Shells pointing out the good and poor aspects of them. One of the Shells was one involving probabilities. I was reasonably positive about most aspects of this Shell but criticised a number of aspects of it. At that moment I suddenly found out that the designer of that Shell was actually in the audience (Oh dear!). He made some comments about my comments, so I asked him to come forward and explain more. It turned out to be a really good intervention. Interestingly, he agreed with many of my criticisms but explained how and why the design problems had occurred. He even allowed me to incorporate some of his comments in the next presentation of the course.

On another visit to Russia in April 1990 I was invited to Leningrad by Prof Leon Mikulich of the Institute of Control Sciences, Moscow, to the IFIP International Conference on Artificial intelligence. I flew to Moscow and arrived at about midday. I then had to transfer to a plane to Leningrad. Initially I thought this was in the same airport (Sheremetyevo) but it turned out that I had to transfer to another airport some distance away (Domodedovo).

At Domodedovo Airport I think I went to the wrong desk (only for Russians, but no-one told me). After I boarded the plane (a Tupolev 144) it rapidly filled up. I was sat in the front compartment at the back of the section in the middle seat with a Russian girl next to me. Suddenly, a Russian man came on the plane and couldn't find a seat. He seemed to be indicating that I had his seat, but I didn't understand. Eventually it was obvious that the plane was full, but he wouldn't get off! The captain argued with him, but the man wouldn't leave the plane. Eventually the Captain switched off the air conditioning and left the plane!

A fight broke out at the front of the aircraft, and I became increasingly alarmed. Suddenly the plane was surrounded by troops with machine guns, and I feared the worst. An announcement came over the loudspeakers. The only word I recognised was "Politzei". The Russian girl groaned at this and said "no politzei!" and started to cry. I could stand it no longer. I got up and, in a panic, I made for the central exit. To my amazement, all the people from the rear half of the aircraft were on the tarmac (it was very hot in the plane) and I motioned to the passengers at the front to follow me and we all got off the plane.

There seemed to be TV cameras there and the extra passenger was still by himself on the plane. A Russian Professor was near me, and he could speak English. I asked the girl (through him) why she groaned when the police were mentioned. She laughed and said that she was making an overnight visit to Leningrad to see her boyfriend, but if the police became involved absolutely nothing would be done, and we would be seriously delayed. Eventually they came and took all foreigners and our baggage back to the terminal (the Russians stayed on the tarmac). The troops went away.

At about 8pm we were taken to board another aircraft next to the original one (the Russians were still on the tarmac). It then became clear that they were having problems with our second plane! Meanwhile the Russians next door suddenly boarded their plane and took off for Leningrad. We waited on the plane in the dark until about 11pm. Finally, the engines started, and we took off for Leningrad, arriving about 1am. Apparently the whole incident was on Russian TV!

A further incident happened when I tried to leave Moscow after that conference. I was required, as Chairman of a UK Defence Committee, to be at a Defence Conference in Helsinki the next day. I was taken to Leningrad Airport by two Russian professors to leave for Helsinki, but when I tried to go through passport control, I suddenly realised I didn't have my passport! It dawned on me that I had forgotten to retrieve it from the hotel desk. The Russian Security Guard said "SO you don't have your passport? I cannot let you through. I hope you like Borscht, you are going to eat plenty of it!" I returned to the Professors who immediately marched into the Airport Controllers Office and demanded to use his telephone. To my amazement he agreed, and they rang the hotel. The Passport was put in a taxi, but it arrived too late for me to board the plane. The next flight was in the morning. The Professors took me back to the hotel and arranged a room.

Then we all went to a traditional Russian Restaurant for a meal. Of course, there was plenty of Vodka and Beer. I remember standing up repeatedly clicking my heels and sinking firstly the Beer and then the Vodka. Things get a bit hazy after that. However, I remember insisting on paying for the drinks and meal. My hosts initially refused "You are a guest of the Soviet Socialist Union" they said. I then pulled out my roubles and said, "I have all these roubles and I need to spend them."

"Put them away," they cried. "That is a lot of money, and you might get robbed—and it might be by us!" In the end I did pay. I seem to remember it cost 22 roubles—for three 3-course meals, lots of beer, Vodka and a bottle of Georgian Sherry! Even at the official exchange rate, it was only a few pounds.

The next day, I still had most of my roubles. At the airport I took them to the travel centre, and they exchanged them for pounds. My hosts were amazed. "That's never happened before!" I was a day late for the Defence Conference in Helsinki and later flew to Glasgow from Helsinki. However, when I returned home my wife Mary said that MI5 had been in contact asking where I was and why I hadn't arrived at Helsinki. I think they were worried that I had been abducted!

Readers might be interested in a presentation given by my ex-head of Department Prof James Cassels, who ended a conference on Nuclear Physics with a brief presentation on how to present slides. The first slide, when shown, appeared inverted. The Prof then remonstrated with the projectionist who reversed it. However, it still appeared wrong. It turned out to be a spoof slide which appeared wrong whichever way it was shown!

Further spoof slides were shown (I cannot remember the detail), but the piece-de-resistance was the final slide. It was a simple graph with two axes. There was a line going up at 45 degrees and a series of points scattered BELOW the line, not fitting the line at all. As the Prof talked (about the good fit) the points slowly moved up to the line, and then all slid down it towards the origin! The line was actually a solid line on the slide, and the points had been stuck on with wax, so as the slide heated up, the wax melted, and the points slowly moved upwards (remember the slide is inverted in the projector) until they hit the line then slid down it! This brought the house down.

So, remember, if you are compromised as you give a presentation, do not give a hint that anything is wrong and carry on as if everything is normal!

Chapter 20
Process Control Research: Combining AI and HCI

One advantage of being both the Executive Director of the Institute and Head of the Scottish HCI Centre was that I could draw upon the expertise of both of my research groups. This happened in two large research projects which we carried out within the ESPRIT programme; a programme announced in 1984 as a European Wide Research Initiative. This involved a huge amount of research money spread across the EU, encouraging industrialists and academics to work together to solve major IT problems and help European Industry.

European cooperation in research can be very successful or disappointing. It depends crucially upon the partners brought together and their capabilities. I have experienced both successful projects and disappointing projects. The first project I was involved in turned out to be a great success. It was called GRADIENT and it brought together both Artificial Intelligence workers and HCI workers (1985—1991). The GRADIENT acronym stood for **GRA**phical **I**ntelligent **DI**alogue **EN**vironmen**T**. Even the title shows the interplay between HCI and AI.

Process Control involves the use of computers to control complex systems which are continuously running. Examples include power plants, industrial production systems, and the systems which control aeroplanes, trains and ships. The problem with such systems is that, in contrast to normal computer programs, they are very complex and continue to operate whatever the operators do (except shutdown which is very expensive). This is totally different from normal computer programs which operate entirely within control of the user. Consequently, such systems must be continuously monitored, and urgent action is often required if the system being controlled moves out of its normal operating conditions. (e.g. an aeroplane going out of control!). They operate in the real-world and can easily move into critical situations either because an internal fault, or some external event, causes problems which need immediate operator correction. If such corrective action is not taken, the damage can be very severe both to the plant being controlled and the environment. An example control room is shown right.

From Wikimedia Commons, the free media repository

Large plants are very complex, often with over 10,000 variables, and the investment in Power Plants, for example, is huge, so any mistakes can be very expensive to correct in addition to being dangerous. Just shutting down and restarting can be hugely expensive. Since monitoring the activity of such industrial plants is the responsibility of the operators, it is imperative that the operators can readily recognise critical situations and apply the correct stabilising procedures. The design of the interface (i.e. what the system tells the operators, and how they respond) is critical to success. Human Computer Interaction (i.e. the way in which the operator interacts with the process system) is therefore a key factor in solving Process Control Problems, and the incorporation of AI in the HCI solutions can enhance their effectiveness.

In some cases, the operators may have little time to correct errors—for example in the control of aircraft. When the flight becomes abnormal, the pilots must act quickly to restore stability. Misunderstanding the situation and applying the incorrect procedures can lead to serious consequences.

Even in the late 80's there were Computer Scientists who still did not understand the significance of HCI Research. The hostility to HCI remained in some quarters even into the 1990's. I remember discussing the 1989 British Midland Crash at East Midlands Airport with a Computer Scientist at Loughborough University. The accident occurred because the one of the engines was failing but the pilot shut off the wrong engine by mistake. The mistake was only realised as the plane made its final approach to East Midlands. The crew tried to restart the good engine, but it was too late, the plane crashed on the M1 motorway, and many people were killed including some car drivers on the motorway. The pilot had shut off the wrong engine because he had misinterpreted the graphical computer interface in the cockpit. I argued that the accident was caused by bad interface design, but my colleague insisted that better engine design was the solution. I replied, pointing out that engine failure was extremely rare and that the plane, in any event, could land on one engine, so the cause was interface design not engine failure. However, he could not see the argument!

There had already been several serious accidents in Process Control caused by operators misunderstanding what the computer interface was trying to tell them. One of the worst examples of these was the Three-mile Island accident, which occurred in 1979 and is the worst accident in the USA's commercial nuclear power plant history.

The problem began with the sticking of a pilot-operated relief valve in the primary system of a nuclear reactor, which allowed large amounts of nuclear reactor coolant to escape. The operators did not recognise this as a loss-of-coolant accident, due to inadequate training and an inadequate computer interface. There had been a series of HCI design oversights resulting in ambiguous control room indicators in the power plant's user interface. An important indicator light which should have been on the interface was not displayed and an operator manually overrode the automatic emergency cooling system of the reactor, mistakenly believing that there was too much coolant water present in the reactor, and that this had caused the steam pressure release. It was a very serious accident and resulted in much more effort being put into good interface design.

In March 1984, shortly after the ESPRIT initiative (an initiative to bring together researchers in Europe to solve important computing problems) had been announced, I met Jens Langeland, from Christian Rovsing in Denmark. He suggested that we might set up a joint bid (with other European colleagues) for a large ESPRIT grant investigation in to the use of Human Computer Interaction in Industrial Process Control. He pointed out that current interfaces to large Process Control Systems were not human-centred and that changes were urgently necessary.

Jens came to see me in Glasgow in May 1984 and introduced me to Peter Elzer of ASEA Brown Boveri (ABB) who was a Process Control Engineer keen to take part in the project. Peter was important because he brought key engineering expertise into the proposed project. Peter could seemingly be quite difficult at times arguing a point at great length and, early in the project, we sometimes became irritated by this. However, it became clear that if Peter had a point, it was usually an important one, although, because of occasional language difficulties, we didn't always understand what he was driving at!

Another participant was Prof Gunnar Johanssen from Kassel University. He was very useful because of his connections with the Engineering Process Control World. At that time, Engineers and Computer Scientists were working on HCI problems independently and were not talking to each other. There was no hostility between the two groups, but their research fields just didn't overlap. Neither group knew of the existence of the other.

At the end of the summer in 1984, we submitted a large-scale proposal to ESPRIT where we mixed ideas from Artificial Intelligence and HCI to create a new joint Engineering/Computer Science approach to process control problems. The ESPRIT evaluating committee liked our ideas but thought that they were rather too advanced, so they commissioned us to carry out a feasibility study (called P600) to test out our proposals and to write a report of our findings.

Nine months later we produced the *P600 Pilot Phase Report Literature and User Survey of Issues related to Man-Machine Interfaces for Supervisory and Control Systems*. The report was published in 1985 and was very well received both by ESPRIT and by the Process Control Community as a whole. It was a large report which we sold for £30, and over the next two years we eventually sold over 300 copies to industrialists across the world.

The report pointed out that many operators were frustrated because the interfaces to the control systems were based upon mathematical models of the process and did not properly take into account the human side of the interaction. There was a mismatch between how operators selected information and how the tasks were to be controlled. Furthermore, computing had taken a number of giant leaps forward in storage capacity and display technology, but these had not really been incorporated into the interfaces. Because of the complexity of the processes some of the information presented to the operators was very difficult to understand.

One serious concern was that as the systems being controlled became more complex, the operators knew less and less about the process itself. Although mathematical approaches to the control problem can be formulated, this often led to difficulties in the human-machine interface because the operators could not form a clear model of what was going on in the process.

In the process control area, it is often difficult to define precise models which account for all behaviour and this had led to a gap between theory and practice.

In situations where a precise model was not possible, attention had been focused on control itself rather than on the process itself.

There was another interesting conclusion from the study. We noticed that as the operations of plants became more and more automated, operator understanding of the processes consequently decreased. In life you tend to learn quickly from your mistakes. As systems ran more and more without operator intervention, operators became less and less competent at dealing with problems because the problems became rarer. It is a bit like having an old unreliable car compared with a new one. When I was a student, I had an old car which often went wrong, but at least I knew how to deal with problems. When I bought a new car, if something went wrong what was happening was hidden under the bonnet and I was often puzzled as to what to do. There were lots of fancy dials and knobs, but I could not get at the actual process going on!

Most of the knowledge which an operator gains about the process is usually acquired through experiencing trouble, but the more an interface is automated the more it often hides what is really happening in the dynamic system, leading to a rarity of incidents and a loss of operator experiential knowledge. One way to overcome this is to provide the full value range of the dynamic interaction parameters, but there are too many, and this often results in the operators being overloaded and confused.

Our Pilot Phase Report had been warmly welcomed by the ESPRIT reviewers, and we submitted a revised version of the full project in 1985. It turned out to be the most successful EU project that I took part in. On 5 August 1985, the ESPRIT Commission accepted the report and awarded us a large grant to carry out research into Industrial Process Control Interfaces over the next five years and the research project GRADIENT (P857) was born.

The set of partners for the project was perfect—ASEA Brown Boveri (a large German Process Control manufacturer) Christian Rovsing (an important Danish Expert Systems company), Interface knowledge from ourselves at The Scottish HCI Centre, Artificial Intelligence assistance from the Turing Institute, and Engineering expertise from the Man-Machine Interface Group at Kassel University (Process Control) and the Chemical Engineering Department at Leuven University.

Thus, began a most successful collaboration which continued for five years (1985—1990). It was a large grant bringing in £480,000 to my research group at

Strathclyde University and I had letters of congratulation from the Chairman of the University Court and the Principal.

I had an amusing experience one evening on returning to Glasgow after a day in London talking to all the participants. I was too late for the last flight, so I booked onto the "Night Sleeper" train to Glasgow which left at about 10pm from Euston. I booked into a first-class carriage and decided to call the guard so I could have a drink. The guard knocked on my door and I opened it but found that the person next door was also at the door—we had both called for a drink at the same time! As I ordered my drink, the person next door said, "Why don't you join us for a drink?" and I did so. He was a tall person with a small beard and his friend had joined us from a nearby sleeper cabin.

Well, once we had started the drink flowed freely and we had a really good laugh. As we passed the station at Crewe, I looked at the person with a beard and I said, "I have a feeling I have seen you somewhere—do you play for Celtic?" He collapsed in hysterics on the bed and his friend couldn't contain himself. His friend said, "Don't you know who this is—he is quite well known?"

"No," I replied, "But he looks like a footballer." This brought even more laughter. "My name is Billy Connolly" the man with the beard said, "I am supposed to be a comedian! —but you're an even better one than me!" I immediately apologised but he said, "Don't apologise, that's the best laugh I have had for weeks!" I honestly did not know who he was, but the next week I actually saw him on television, and it turned out that he was quite famous.

In early Process Control Systems, the dialogue (i.e. how the system communicated with the operator) was intimately encased within the control program. It was fixed and non-variable. We felt that different situations would require different presentations of information, for example, in some cases a simple text message might suffice, but at other times a video of part of the plant combined with contemporaneously observing some variables changing with time on a graph, might be needed.

We therefore suggested that the communication with the operators should be separated out from the main control program and handled by separate software called the Dialogue and Presentation Systems. The Dialogue System is concerned with *what is communicated and when* (i.e. rise in temperature, too much or too little cooling, stuck valve, input, and output issues etc.) and the Presentation System determines *how the information is presented or collected together* (should it be by text, graphics, voice output, and how much information should be communicated). This was the HCI side of the Project. The diagram below shows the idea.

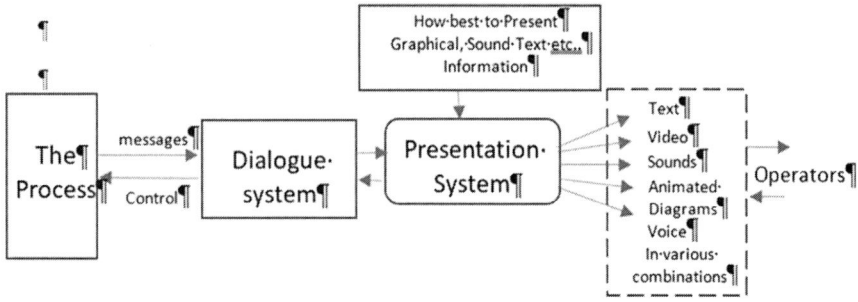

Messages from the process generate a dialogue with the operators. The Presentation System then decides how best to present the information to the operators using a combination of presentation techniques and appropriate output devices. Of course, access to Process Knowledge is required (what is actually happening in the system, and what procedures can be used to correct outages). This was the Artificial Intelligence contribution to the project. The Dialogue and Presentation Systems were therefore supported by three distinct Expert Systems (to remind yourself of what an Expert System is, please see page 142).

Early on in the project, Peter Elzer had suggested an Expert System he called **QRES** (Quick Response Expert System). He argued with us at length for a couple of meetings over this particular issue and at first, we didn't understand what he was driving at! After long discussions we finally understood. He was pointing out that in large industrial process control systems, when something goes wrong, the most important thing to do is to bring the system *into equilibrium* in order to give the operators and the engineers time to sort out the problem. Once the system is in equilibrium the operators can relax and determine what is wrong. He therefore proposed the construction of the Quick Response Expert System (QRES) to do this function—to buy time for the operator to bring the system back to normal. This Expert System identified single system failures and failure states and evaluated their severity and then recommended predefined stabilising actions. There were a number of other expert systems which understood the process, but I won't go into detail about these.

The Dialogue System ran a series of Dialogue Assistants to handle various situations. The dialogue was then interpreted by the Presentation System, which decided exactly how the information was to be displayed to the operators. One novel approach in the project was the construction of a Graphical Expert System (**GES**) containing the pictures and picture sequences from which to choose the right level of interaction. Depending upon the situation and how the operators were handling the problem it could adjust the actual presentation of the information. So, the system could present alternative representations to assist the operator. Alternative media (text, text with diagrams, voice with diagrams, video etc.) could be used to improve operator understanding.

This proposal for flexible on-line picture creation was very unusual at the time. In conventional power stations, the basis for the design of the "control room pictures" often came from pictures from a previous plant, or a standard set which had been developed over the years. Such designs contained a lot of "implicit" knowledge. To create these, the Engineers would have had extensive discussions with Picture Design Specialists in a very time-consuming process. However, the way in which picture objects are combined was fixed at design time, which meant that the system could not adapt to developing situations.

In our approach, the collection of graphical objects, text and other media, could be combined in different ways when the process was running, and at different levels of detail. It was therefore possible to switch between different levels of detail, and to combine, on-the-fly, different graphical objects to display the actual process variables. Operators could also ask for particular types of presentation.

During the running of the GRADIENT project, it was realised that the use of many different media in computer systems was becoming increasingly important particularly as the cost of storage was rapidly decreasing. This allowed us to enhance the output to include tonal sound, synthesised speech, graphical interfaces, 2- and 3-D graphics, animation and live and pre-recorded video.

Another important concept was what we called "Unmediated video and audio." This meant using data in unusual ways. I thought of this idea from my early experience with the Liverpool University 1906S in the 1970's. The Engineers had attached a loudspeaker to the computer console for their debugging purposes and it was for their use only. However, the speaker was left connected after the computer was accepted by the customer. Such information was not intended for operator use, but the operators could turn up the volume and listen to it. Surprisingly it became a useful operator aid. The connection between the sounds emitted and what was happening in the machine was never explained, but the operators were rapidly able to exploit it. This is what we meant by the term "unmediated".

The final review of GRADIENT took place in 1990, as I left the University of Strathclyde and took up the position of Prof of Computer Science at Loughborough University. GRADIENT was the most successful EU project that I took part in and was certainly one of the most successful projects in ESPRIT. It had just the right combination of academic and industrial collaborators. I think it had an impact on future interface design in large Process Control systems.

The final project reviewers were Prof Morton Lind and Dr Thies Wittig. The day before the review, I was walking down the street in Amsterdam when a man on a bicycle nearly ran me down. He profusely apologised and it turned out to be one of the reviewers. "Is that what you think of the project" I said!

At the final review, on 4 October 1990, the reviewers praised the project and were convinced it had, and would, affect the design of Process Control Interfaces in the future. Some important Research Papers were produced. It introduced the combination of HCI research and AI research into Process Control Interface design. It was generally agreed that the project had a major influence on Process Control Interfaces.

Just before I left Strathclyde, Prof Walter Bogaerts, who worked with us on GRADIENT, asked me to go to New Orleans to address a Materials Science conference on Expert Systems (of which they knew little). I went there for three days and delivered a series of seminars. They then invited me to organise their next seminar in Scotland in November 1989.

I organised this at Lomond Castle (a beautiful location on the south of Loch Lomond). Amongst the attendees was a Dr Rene De Planque from the Fiz-Chemie Institute in Berlin (an organisation which produces the Handbook of Chemistry). Rene and I got on very well. I remember pointing out to him by the loch one evening that I could walk on water!

We were by the lake, and it was dark in the moonlight. I illustrated this by calmly walking about 20 yards out on top of the lake surface. The water was just about covering my shoes—and they were astonished. Of course, I hadn't told them that the loch was unusually full and a wooden jetty, which projected out about 50 yards into the loch was just submerged by about two inches of water! They finally realised what was happening and ALL of us walked on water. Rene and I became good friends and colleagues and later, we worked together on a large important German project to assist student learning in Chemistry.

The work we carried out in the GRADIENT project, had suggested that we ought to think about the potential of using much richer media to communicate information in process control. In 1989 a second European initiative had been announced (called ESPRIT-2) and we decided to bid for another Process Control Project. From our extensive work on GRADIENT, we have been able to gather together some guidelines or principles for designers. Although the process control area is rather specialised, we are convinced that the lessons we had learned from this work were readily generalisable and were therefore applicable to most multimedia design situations.

Chapter 21
Loughborough: Multi-Media and Process Control

On 1 October 1990, I joined Prof Ernest Edmonds at the Computer Science Department at Loughborough University, to take up the post of Professor of Computer Science. I left Strathclyde University and Turing Institute with some sadness since I had really enjoyed my time there. In particular, I was very grateful to both Sir Graham Hills (The Principal of Strathclyde) and Lord Balfour (Chairman, Turing Institute) for all the help and encouragement they had given me. Lord Balfour insisted that I remained on the Turing Board as a Board Member.

As I moved to Loughborough I was invited to my old School, King Edward VII, Lytham, to be the speaker at the Speech Day. This really was a surprise. I had not been particularly well thought of at School and was surprised they even remembered me!

Back in 1956, when I was in the sixth form, a pupil called John Tagg (who later joined the Air Force) and I, carried out a prank on the girls' school next door (Queen Mary's School). It was the year of the Hungarian Revolution and after dark we climbed onto the Dining Room roof and scrawled on the roof (in washable white paint) "Hungary Forever"! We then rang the press and a half size picture appeared on the front page of the local newspaper the next day. We never told anyone we had done it and assumed that no-one knew.

At the Speech Day, as the headmaster introduced me to the parents and the boys he said. "This is the man who wrote "Hungary Forever" on the Girls Dining Room roof, and I think it got the girls some extra rations!" We were convinced at the time that no-one knew who had done it! I gave an address to the boys mainly about the Turing Institute and Artificial Intelligence and asked for the usual half-day holiday (though I doubt if they received one!).

When I arrived at Loughborough, I decided to bid for a Process Control project in which we explored the possibilities of Multimedia interfaces (interfaces which use a variety of media—text, voice, videos, animation, sound etc.). Our partners were Tecsiel (an Italian software house), Scottish Power (the Prime Contractor), DOW Chemicals (Netherlands), the University of Leuven (who had worked with us on GRADIENT), University College (Dublin), and the Work Research Centre (Ireland). The Title we gave the project was PROMISE which stood for **PR**ocess **O**perators **M**ultimedia **I**ntelligent **S**upport **E**nvironment.

What made it particularly interesting were the two commercial operators (Scottish Power and DOW) who could provide live and substantial applications (a nuclear power station and a large chemical plant). The Work Research Centre from Ireland offered expertise in practical experiments which would be needed in the project. The project was approved for 4 years (1989—1993). My group received £700,000 to fund the research (initially at Strathclyde and then later transferred to Loughborough University).

Most of the Research Assistants moved down with me. In PROMISE we were particularly interested in how multimedia interfaces might improve the effectiveness of the operator-system interface.

What do we mean by "Multimedia"? The media are mediums of communication. There are INPUT media with which the user tells the computer what to do. Then there are OUTPUT media through which the computer communicates to the operator. Examples of INPUT media include typing text, voice input, and using a mouse to highlight something on a computer screen. OUTPUT media include print-out voice-output, sounds, music, pictures, diagrams, and videos. Most computer dialogues in the 1970s only used text to communicate with users. Then, in the 80's came along graphics and windows. Voice Output also began to be used in the 80s. As storage costs plummeted, video also became a possibility in the 90s. By adopting these new technologies, enriching the communication between Operator and Process became a possibility. But the key questions, as we saw it, were:

1. When are certain media combinations more appropriate than others? What are appropriate combinations of media which will have the maximum effect? For example, if information was presented with diagrams and speech, would the operators understand the message better than if diagrams and text were used?

2. Are some combinations of media less efficient than others in certain situations? Do some media combinations result in better learning and understanding?
3. Is the effect of a medium or combination of media different depending on who you are? Do some individuals prefer text messages whilst others prefer visual messages or spoken messages?

The literature was searched and there had been some tentative explorations, but no-one seemed sure how to characterise media.

Usage of multiple media involving, say, ears, eyes, and both hands, could enable the user to perform several tasks at the same time and several users to perform the same task at different times. In other words, multimedia interfaces could be viewed as a multisensory, multichannel, multitasking and multiuser approach to systems design (that's a mouthful!)

Up to 1989, the development of computer applications had often been seriously hindered by technological approaches. Very often designers had worked outwards from the technology asking, 'what might the user *be able to do* with this new technology?' rather than 'what might the user *want to do* with this new technology?' In many ways it was the same mistake as happened in the design of the first computer interfaces back in the 1970's!

Many developments in the multimedia area illustrated this point. Designers often start with the concept of the linking of many different types of media ('Hypermedia') together. Technologically this is certainly possible. They then suggest new ways of improving the education of users using these highly linked systems. Whether users actually want to do this often is not questioned. Because of such approaches, by 1989, a set of multimedia (or hypermedia) applications had been created which were of dubious value.

The emphasis on these rather artificial explorational and educational aspects of multimedia obscured other, perhaps more important applications. Although human beings do not, as a rule, get urgent desires to examine high-definition pictures or watch snippets of video whilst reading books, they do frequently exploit multimedia aspects to improve understanding of the world around them. For example, it is well known that using both visual and audio channels simultaneously to explain a complex diagram is better than using only one channel.

It is also true that human beings use the apparent redundancy offered by multiple channels to improve their understanding of situations, an example being the use of gesture, audio and visual cues whilst taking part in a multi-user conversation. The importance of such cues is illustrated by the fact that tape recordings taken of what were apparently perfectly understandable group meetings turn out to be virtually unintelligible when played without the visual cues, even though the speech content was the most important.

The other important guiding principle of the PROMISE project was the importance of working in the real world. Our work was therefore based upon a nuclear power plant simulator in the South of Scotland (Scottish Power) and a chemical plant in The Netherlands (DOW Chemicals). These two plants were deliberately chosen for the different environments they provided. Work on the nuclear plant was carried out in the plant simulator, one of the most sophisticated in the field. The work in the chemical plant was carried out in the actual control room itself. The simulator allowed us to deliberately create error situations and explore the importance of different media in assisting operators to deal with these situations.

A real plant (see example right) offered quite a different challenge—that of studies in a real environment where the clock cannot be stopped, and unexpected events are intermingled with long periods of trouble-free running.

In addition to experiments in the simulator and nuclear plant, parallel experiments were also being carried out in the Behaviour Laboratory at Loughborough University where conditions could be carefully controlled.

Over the four years of the PROMISE project, we identified a number of useful principles for designers concerned with the creation of interfaces which operators could use.

1. Connection with the Real World.

We found that Multimedia options such as video and audio allowed *operators to regain something which they had lost over the previous twenty years* of process control interface design—that is—connection with the real world! In days gone by, the operators could FEEL the plant under their feet, HEAR it in operation, or SEE what was happening and in some cases, SMELL the overheating. Such events had been largely replaced by proxy events. A proxy event is an event substituted for the real-world event. So, a rise in temperature in the real-world causes heat to be produced, but the change in a dial in the control room is a proxy event. The real-world option offered by a multimedia approach partially eliminates events by proxy and preserves crucial cues for the operators.

2. The Dangers of Events by Proxy

These *events by proxy* (for example, a red light could change to a green light on the control panel to signify the opening of a valve) do not necessarily meant that the valve has actually opened. Events by proxy can be a serious source of problems in process control. Using a multimedia approach, the operator could actually watch the valve changing state, and hear the valve open.

I personally experienced an example of the importance of real events back in 1972. I had applied to IBM to become a Systems Engineer and I flew down to London for the interview. After the interview I flew back from London to Liverpool. In those days the aircraft was a Vickers Viscount and the last flight of the day landed also at Chester but then it would do the 15-minute hop over the Mersey to Liverpool. The flight was very calm until we approached Chester but then it became quite bumpy. Suddenly the co-pilot came out of the cockpit and came down the plane. He asked me to move out of my seat and he then shone a torch out of the window onto the wing. Planes (even today) have a small, raised plastic window on the upper side of the wing and when the undercarriage is down a small metal bar can be seen (which can't be seen when the undercarriage is up). The pilot was checking if the under carriage was really <u>down</u>, because the indicator light in the cockpit was showing that it wasn't!

The bar on the wing is a real event. The light in the cockpit is an event by proxy. The plane circled the control tower at Chester and asked them if they could see that the undercarriage was down, but it was very dark. The pilot eventually announced that there might be a problem with the undercarriage, but they were not sure, so we were abandoning Chester and flying to Liverpool which had a much longer runway and better emergency facilities. The pilot said that he thought all was Okay (because of seeing the real event on the wing),but he had to be careful.

We came into land at Liverpool, and we could see the fire engines lining up by the side of the runway! However, the landing was uneventful. The non-proxy event was correct.

3. Which Cues are Really Important?

It is often difficult to be precise about exactly what in our environment provides the important cues for human decision making. Media (particularly video and audio) carry information which cannot be easily characterised as having a particular use. Good operators simply "know" that there is a problem when they hear a peculiar sound from the plant, yet they may be unable to define what it is that alerts them. In a similar way the average car driver can sense there is something wrong with a car simply because of a change in the sounds produced whilst running but may not know what precisely is causing the unusual noise. Operators use cues which designers do not always know exist (and in many cases the operators are not conscious of using these cues either). Designers tamper with such information at their peril.

In one of our experiments in the chemical plant, for example, we found that there was a problem when using the results of a test involving an assessment of the salt level in the process flow. Designers had assumed that a measure of the length of the deposited salt in a narrow tube would be sufficient, whereas it turned out that the colour of the substrate, and even the granulation of the salt, were also used in an important way by operators in the evaluation process. Preserving the complete test in a still-colour-picture preserved the vital cues and improved decision making.

4. Changes to Media need to be large enough to be recognised

In several of our laboratory experiments we provided additional sound from the process to help the operators make their decisions, but in some cases such sounds were found to be irritating and not very useful because the operators could not use the information in a meaningful way. The differences in the medium were not translated into quantifiable differences for the operators to appreciate, thus control was not possible. An example of this would be representing flow by the sound of water from a tap. Increases or decreases in sound would be perceived but absolute measurement is not possible. Thus, the use of such a medium would only be appropriate when gross changes were the major cue for action rather than precise changes.

We concluded that the most important factor in deciding which media to choose was the use to which the information is being put by the recipient. People will argue that this cannot be true. Surely, spatial information is always better presented pictorially? This is not necessarily so. For example, experiments carried out with weather forecasts have showed that audio information can be superior to visual information in some circumstances even in what appears to be a spatial task.

The information needs of users will, of course depend upon their goals, but it will eventually need to be converted into requirements for particular types of information. For example, operators watching a boiler heating up may have an overall goal of preventing the temperature from rising too quickly. They will therefore be interested not only in absolute temperature values, but in rates of change as well. They might also be interested in the history of the previous five minutes. Alternatively, if rates of change were unimportant, they may only need to know when a temperature level is reached. In each of these cases the information required has very different characteristics. On the one hand, values, rates of change and historical perspectives are required, but on the other merely a change of context. It is clear that using an audio medium for the former task would be inappropriate, but might it be ideal for the second case.

5. *Is Using Multiple Media really useful for Complex Situations?*

A question which had never really been answered was "can a multimedia presentation approach assist the user in understanding a complex situation?" Most people would feel that the answer ought to be yes, but there was little data to support this viewpoint. We therefore carried out a set of laboratory experiments to investigate (amongst other things) how far this might be true.

One conclusion was that if a concept is easy to understand, almost any medium combination will be successful. In contrast, if a concept is really complex then no medium combination will get the idea across unless the operator understands the situation. However, it is in the middle ground that differences in comprehension occurs. Where there is incomplete user understanding, some media combinations will work better than others.

6. More is Better

We eventually decided that in the multimedia situation—more is better. This may seem a strange decision. Surely too much information could confuse? However, human beings seem to prefer parallel sets of redundant information rather than unique sets of accurate information. We are using the term redundant to mean "useful but not strictly necessary". It appears that the more human beings use redundant (but individually useful) information the nearer they get to exploiting the "whole" mind.

An overhead colour shot of a snooker player sinking a shot is enough to tell an observer what is going on, however, the sound of the ball dropping into the pocket, although apparently redundant, does aid comprehension. It gives an alternative, reassuring confirmation of the event. This is an important use of multiple media.

7. Intrusive Media and High-Level Goals

Some media are more powerful than others at helping operators to solve certain problems. An example of an intrusive medium is sound when used to alert operators of alarm conditions. Such media can be very successful but can impair performance on other parallel goals. Therefore, intrusive media should only be used when the intrusion is required to solve a higher-level problem and when the operator is prepared to risk the loss of concentration on other lower-level problems (and designers must remember to be sure to re-introduce the neglected problems when the emergency is over!).

8. Media Switching

A switch of medium can be very effective in forcing a context switch on an operator. Sound has often been used for this purpose (few people can ignore a sound alarm), but other media switches can work as well. For example, a visual switch in a sound intensive environment can be very effective (provided the operator's attention to the screen can be guaranteed). Media switching can also shed new light in problem solving.

Some media work extremely well if synchronised together. The obvious example is of video and audio in film or television and the synchronisation which carries over a notion of reality.

PROMISE was an interesting project and was well received at the final review in 1993 and we hoped that it would have an impact on new interface design. However, because Process Control Interfaces are really complex, and take a considerable time to implement, it was a few years before designers used some of our ideas.

Shortly after I arrived at Loughborough a strange thing happened to me—I had an out-of-body experience! In March 1994 I went skiing with my wife Mary to Switzerland. Whist there I suddenly had serious pains in my chest and thought it might be a heart attack. However, after about half-an-hour they suddenly disappeared, and I assumed it was indigestion. Later, it happened again in my office at Loughborough. The pain was really bad and came on suddenly. My secretary called the medics, and I was on the floor. However, as they arrived the pain suddenly went completely, and I felt fine. The doctors examined me and said there was nothing wrong with my heart, but I ought to have investigations since it seemed to be connected with digestion.

Over the next few weeks, it happened a number of times and the doctors assumed it was irritation of the oesophagus. But on a Sunday in September, it happened again but the pain didn't go away. Eventually, Mary called an ambulance, and I was taken to hospital. I was violently sick in every department of the hospital, and finally at 11pm the doctor came and said, "We know exactly what's wrong with you, do you want the good news or the bad news!"

"Give me the good news first" I replied. "The good news is that we know that you have Acute Pancreatitis and know how to treat it. The bad news is that it will be very painful indeed for the next 10 days!" For the next ten days I was in hospital "Nil by mouth" (not even water). I was kept alive on a drip.

The Consultant said that the intense pain would remain for 10 days. They would give me powerful painkillers, but they would only last for three hours, I would have to manage the final hour with a progressively increasing pain. This I did for ten days. On the ninth day I was getting really down, and I asked the consultant how long this would go on and he replied, "Until tomorrow, the pain will stop then."

"Are you a betting man?" I replied. "I bet you £10 that it will not go by then."

"Done," he replied. The next morning the pain suddenly stopped as he had predicted. I offered him the £10, but he refused. "I knew it would stop. If I take your £10, it would be like taking money off children!"

The problem was caused by a gall stone. When you start to eat, bile flows down to the intestine from the gall bladder. There it mixes with an enzyme from the pancreas, and it becomes a flesh-dissolving enzyme. The bile tube meets the pancreas tube just before the intestine and there is a one-inch common tube. It was in this tube that the gallstone became stuck. This meant that the bile flowed upwards into the pancreas, the bile and enzyme mixed—and started to dissolve my pancreas! It is very serious, and you can die if it is not treated.

After coming out of hospital the doctors arranged me to go in for a gall bladder removal operation in about 5 weeks. Unfortunately, before that I had another attack and was again rushed into hospital. Mary came in with me, and they immediately decided to administer the very strong pain killer. By the side of the bed was a machine for measuring blood pressure and heart rate and they connected me to it. They then administered the injection.

Almost immediately I felt strange. I could see clearly and hear everything being said, but my body slowly began to move away from the bed (backwards). I could clearly see the doctors talking with my wife, but I couldn't move or say anything. I thought—Am I dying? Well, it is not too bad! No-one initially seemed to notice, but as I appeared to move away from the bed, I suddenly saw total panic break out before me and I heard one doctor exclaim "his blood pressure has dropped to zero" and another said, "His heart has stopped!" I could see and hear all this clearly. At that moment I felt an incredible urge to be sick. I raised myself up and did a projectile vomit which my wife caught in a bucket! Then I collapsed back on the bed and my body slowly began to move back to the bed. As I reached the bed I could move and speak again. The doctor said, "Don't you dare do that again!"

I suddenly felt normal again. I can't explain it—but that is exactly what happened, and I was conscious throughout. I have to admit that I didn't hear the Lord saying, "No Jim it's not your time—go back"! Nor did I hear angels singing. However, it was a very strange experience. I talked to the Doctors afterwards and they agreed with my recollection as to what had happened. The machine had indeed showed my vital functions stopping and Mary had caught the vomit in a bucket! Other people have talked about such experiences, and they are called Out-of-Body events. I can't explain it.

Chapter 22
Composing Music

This chapter provides a little light relief from talking only about Computing and is a precursor to the next chapter on using music in computing. The chapter discusses how people compose music. After reading it you might be able to compose a popular song and make lots of money (or maybe not!).

I have already discussed two events which changed my life when I was 16. In fact, there was a third event and Peter Smith was again accidentally responsible and it had a major impact in my Computing Research later at Loughborough University. Both of us had been playing the piano for a number of years and we were reasonably competent. I had played many classical pieces for the piano, and I enjoyed playing them, but I did not really get a "buzz" out of them.

In 1956, Peter approached me and suggested that "the two grand pianos in the main hall lie idle after school, so why don't we learn to play some pieces for two pianos?" He had in mind some Rawicz and Landauer arrangements (they were a popular piano duo on TV in the sixties and seventies) and we certainly played some of their arrangements at the school concerts. However, one day he brought in a two-piano version of Grieg's Piano Concerto and suggested we had a go at the first movement. "What's a Piano Concerto" I asked. "It is a work for Piano and Orchestra in three movements—something of a showpiece for piano" he replied. "I have the score where the orchestral part is reduced to a piano part, so why don't we play it?" So, Peter (who was a better pianist) played the Solo Piano part and I played the Orchestral Reduction.

I immediately went out and bought a copy of the two-piano version and took it home so that I could learn the orchestral reduction. Over the next few weeks, we progressed quite well in the first movement, and I asked Peter if he had a recording of the work so that I could actually hear the orchestral part. In those days orchestral music was on 12-inch 78 rpm records and the Grieg Piano Concerto was on three double sided records. He therefore let me borrow the first record which contained the first movement. I played it a few times so that I could improve my piano part and at first that is all that happened.

After about the 5th time I played it, I suddenly realised that some bits of the concerto sounded rather good, and I replayed the record. Each time I played this first side of the record, the good bits got better and better and extended, and eventually the whole movement just knocked me out. I couldn't stop playing it! It was my first experience of the real "high" that Classical Music can give you. It was like a drug! We did continue playing the Grieg together, but I became obsessed about the rest of the piece and Peter lent me the second and third movements on 78 records to listen to. I played the second (slow) movement and was disappointed. Poor old Grieg had obviously lost his marbles on that one! I then tried the third movement, and after about three hearings it began to excite me. My experience of the first movement made me persist and it eventually gave me a high as well. Finally, I went back to the second movement and the same thing happened again. He hadn't lost his marbles! Now, playing the whole piece gave me an intense high. To get a "high" from Classical music you have to listen to it (seriously) for about 5 times before it gets to you, but the effort is really worthwhile.

Frank Duckworth, who was also interested in Classical Music, joined us. We studied other works—Rachmaninov's 2nd Piano Concerto, Gershwin's Rhapsody in Blue, Gershwin's Piano Concerto and Ravel's Piano Concerto, all of which, after listening about 5 times, knocked us out! We gradually increased our repertoire and moved into Orchestral Music—Holst's Planet Suite, Gershwin's American in Paris, Debussy's Le Mer and Walton's Belshazzar's Feast. We probably looked an odd set of teenagers, jumping wildly about the room, in ecstasy at the music.

Ever since this experience, Classical Music has dominated my life and given me the greatest pleasure. I repeatedly tell people that you must put in the necessary work (say 5 listenings) before you will get the high result, but you don't get the reward if you listen casually, which most people do. Classical Music is not a polite form of music which people dress up to go and listen to. It is a drug which gives you an intense high, but you must listen to the music about 5 or 6 times before it will grab you! This appreciation of Classical Music eventually had a major influence in my later Computer Science research.

I suspect that the desire and ability to compose music is in your genes. Strangely, even from the age of 12 years I had a desire to compose music. I don't know why but I was continuously on the piano trying to write music but without much success. Once I began to listen to classical music, however, I began to realise how music was composed. At 16 years old my first composition for piano was played at a school concert. It was called "Phantasie No. 2". It was Frank who suggested the title. "Don't let the audience know that it's your first work!"

So how is music composed? The number of possible tunes that can be written is huge. Think of the piano. A composer could start the tune on any one of at least 24 notes (assuming that very high or very low notes are not good starting points). Each succeeding note could be any one of at least 12 notes (again assuming that really large note changes are not desirable). So, tune with a length of say, 24 notes will have 12^{23} different possibilities! No wonder we don't run out of tunes to write!

There are ways in which this number of possibilities can be reduced. One technique developed in the 17th Century is the key signature. An example of a key signature is playing the white notes only on the piano starting at C (called C Major). The tune often starts at C and nearly always ends at C, and the number of notes is restricted to 7 of the 12 available (in C major, the black notes are not allowed). This reduces the number of possibilities to about 8^{23}. There are other constraints which further limit the acceptable possibilities. However, human beings like change. A tune consisting of just two notes repeated 24 times would be regarded as boring. Likewise, a tune in which the notes leapt about all over the place would be changing too much to be acceptable. The composer treads a fine line between being boring and being too complex.

I began to realise that composing music was very much connected to Mathematics, Computer Science and Psychology, when I attended an Open University Psychology Course in 1978. In the early days I would create a few bars and then not know what to do next. But after appreciating the works of classical masters and learning about memory in Psychology, it suddenly hit me: it has a lot to do with short-term memory!

Human beings have a short-term memory and a long-term memory. Long-term memory is where we store recallable memories. We tend to use short-term memory when we are listening to something new. Short-term memory can only store about seven items (or strictly speaking, what are called "chunks"). So, if someone says a series of number to you, you will fail to remember them properly if the number of items exceeds about seven.

If I say to you the number 1,5,7,3,4. You will retain it for a while. However, if I say 2,6,7,3,5,4,8,6,2,4,1,9 you will lose it as I progress through the numbers. You might remember the first two or three numbers and the last two spoken, but you will lose the ones in the middle.

The effect is described in a paper by Miller—*The Magic Number 7 +- 2* (Miller).

I used to show students how valid this was by reciting a series of whole numbers (0 to 9) one after the other with a half second pause in between, and, at the end of the sequence, asked them to immediately write down each number as they had heard them. I would intersperse short lists like 1, 5, 2 4 5, and larger lists like 2 9 6 7 3 8 2 6 7 5 9. At the end of the experiment, I asked how many students had written down the right sequence and we always got a graph like the one below.

Notice that there is almost 100% recall until we reach sequences of length 6.

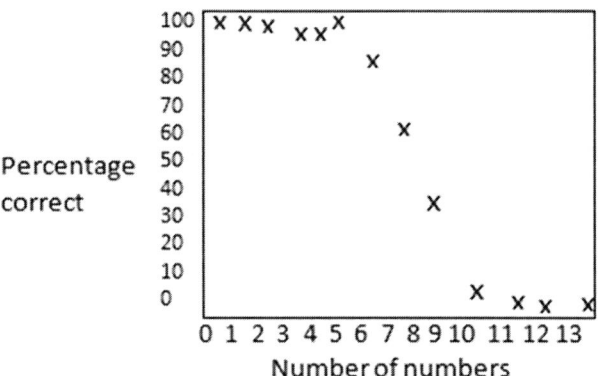

Then recall deteriorates quickly and by length 9 it is effectively zero. I did this for many years and the results are always the same—Miller's graph is a really robust result.

The problem with appreciating music is that unlike written words the listener cannot pause and examine the music. It comes at them in a continuous stream. You hear a few notes and then you must move on to the next few notes. With reading a person can re-read a section which they do not understand.

It seems therefore to me that music must consist of a sequence of very similar short groups of notes (i.e., with a length of about 6 or 7 notes) and a succeeding group must be simply related to the previous group. If not, short-term memory will fail. Of course, we can eventually remember the music in long term memory, but on first few hearing you have to use short term memory.

So, a musical piece begins with a short pattern (6 or 7 notes). Then the pattern is slightly modified (or repeated). Then further modifications take place.

But what sort of small changes would be acceptable? Abo are two octaves

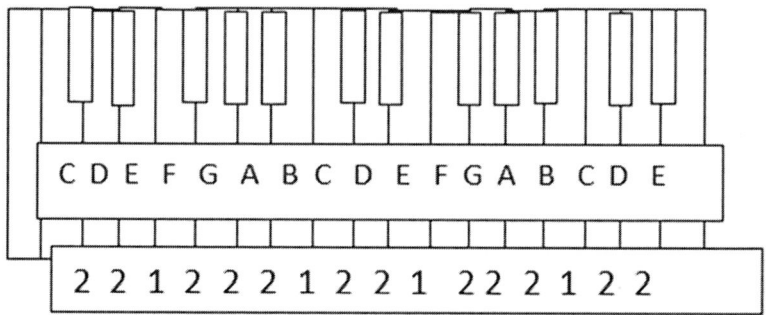

on the piano. The white notes are labelled C, D, E etc, and between some white notes is a black note. Not every two white notes have a black note between them. The distance between each note is called a semitone. In the diagram the number of semitones between each white note is shown (usually two but occasionally one). Notice that the intervals change as you move up the piano. There are two semitones between C and D, and D and E, but only one between E and F, then 2 again between F and G and G and A, then only one between B and C. A black note is an extra semitone between white notes.

So, if CDE is played it sounds slightly different than DEF, and different again for EFG, but FGA and GAB sound similar to CDE (but higher up the scale). This means that the same sequence of notes sounds a bit different if played on different set of white notes. This gives composers the ability to slightly change the note sequence without changing the sequence too much. Short-term memory can cope with that.

Another small variation can be achieved by inverting the pattern or playing it backwards. Composition is therefore about making small changes to the pattern as it progresses so that short-term memory can keep a track of it. Another very important aspect of music is Rhythm. This again keeps sequences sounding similar.

Let us examine a simple tune and look at the pattern changes—God Save the Queen.

***

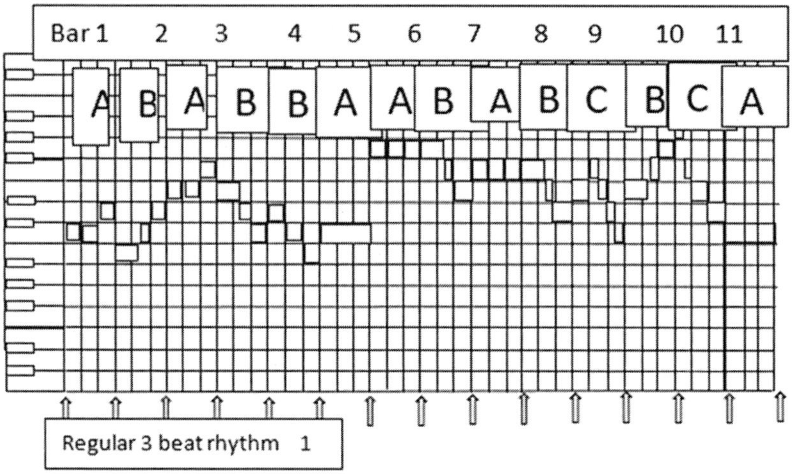

[Figure: Graph showing bars 1–11 with pattern labels A B A B B A A B A B C B C A, piano notes on the side, labeled "Regular 3 beat rhythm 1"]

The graph above shows the piano notes at the side and time progresses along the x axis. It begins with a pattern (A) of three notes—CCD, followed by a second pattern BCD (B). Next, A is repeated two tones higher EEF. This sounds slightly different than CCD. Then the second pattern BCD is inverted two notes higher at EDC (call this B*). Then the pattern B* is repeated but one tone lower DCB and it finishes where it began at C (with a slightly modified pattern A*). Then we go up five notes to G, where we have pattern A*. This is followed by the second pattern B*. We drop a tone and repeat A* and B*. Now we add a few extra notes to pattern B* EFEDC but this is really a slightly modified B*. We repeat B followed by a modified C* and end on A*. The whole tune is based upon two patterns A and B.

Note, the rhythm also helps understanding with a regular three beat time signature. So, music composition is very like Computer Science. It is all about pattern matching and minor changes. The next example is more complicated than God Save the Queen, but it follows the same construction mechanisms.

The diagram for the carol "Silent Night, Holy Night" is shown overleaf.

The piano keyboard is at the base of the diagram. Read it with the page on its side.

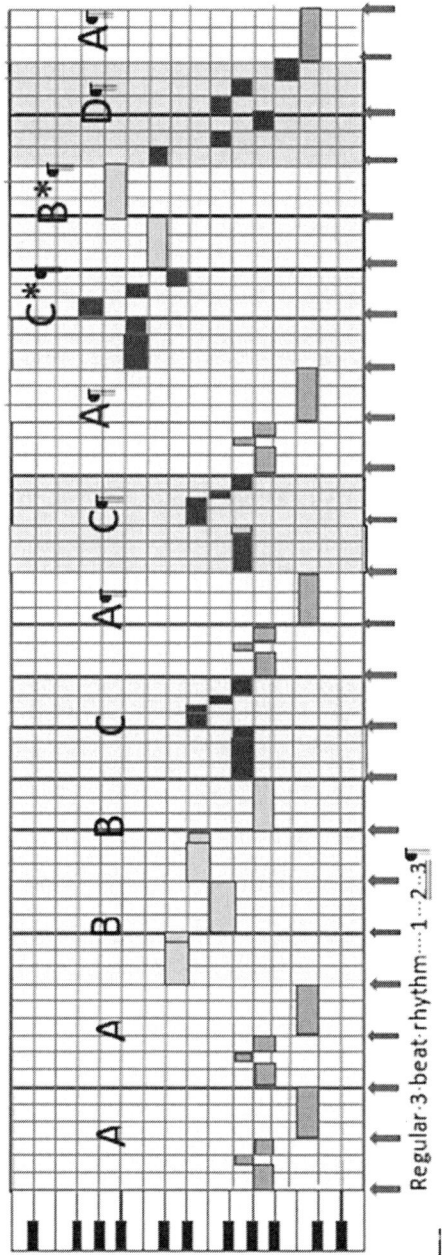

The carol starts with a simple 4 note phrase over 2 bars (labelled A). This is well below the maximum of 7 so short-term memory (STM) will have no problems understanding it. A is then repeated which reinforces it in memory. Then the phrase changes (to B). Notice that B is really the end of Phrase A. with different timing. B is then repeated a little lower. Because it is on a different set of white notes it will sound slightly different. Now C is introduced but this is really a small modification of A, and it is followed by A. These two phrases are simply repeated.

C* is C slightly changed (however it does look very like A!). The song ends with D which is two sets of descending notes, and the carol ends on A.

Note how it is very carefully constructed so that short-term memory can keep track of it. Even the first time the carol is heard the listener can easily follow it.

This carol is one of the most popular in the repertoire, and this is very much due to its careful construction.

The various phrases also are also slightly changed rhythmically as well. A is Daa-DiDi Daa-aa. B is Daa-Di-Daa. C is Daa-Di-Daa-DiDi etc.

Now when you listen to music, pay attention to how the short set of notes at the beginning are slowly and gently modified. Of course, there are a lot more techniques that composers use to write works into various extended forms— Rondo, Variations, Sonata Form etc. A larger work needs to be organised carefully, so the listener does not get lost or bored.

Having realised the importance of short-term memory, one aspect of music composition really troubled me—Atonal Music. At the beginning of this century, Arnold Schonberg proposed a radical change in composition technique. As the nineteenth century was coming to an end, composers became more and more adventurous in the use of Key Changes. Debussy, for example, astonished the musical world with "Prelude a L'Apres Midi du'n Faune" with a set of amazing key changes. Schonberg suggested that composers could throw off the shackles of Key Signature and produce music which was Key-less. To do this he postulated that one set of notes should not take precedence over any other, so he proposed that the music would be based upon tone-rows consisting of twelve notes, with no note being played twice in the row. Many composers have since attempted to write such music (e.g., Berg and Webern) but to be honest it has not really caught on. Some Berg is quite well known but even he breaks the principle occasionally.

In my view, the problem with Atonal Music is that the initial set of twelve notes—the tone row, is too large for short-term memory to assimilate particularly if there is no key signature. The human brain is not designed to follow sets of 12 note phrases and listeners get lost. Of course, if one listens repeatedly enough one can eventually get the sequence into memory, but most people are not prepared to make that effort. I think it might work if the tone row was divided into two and the second half was a minor variation on the first notes of the row.

After I went to Liverpool University, I continued composition. In my second year as an undergraduate. I entered a competition for amateur composers through the University. The judge was the Prof of Music (Gerald Abrahams) who was also the Head of Radio 3 music. There were 8 or 9 entries and I decided to write a short piano work (which I wrote in two days).

The work was in the style of J.S. Bach but with a modern twang (this was a mistake!). I therefore called it "J.L. Bach", J L being my initials. Frank Duckworth played it for me at the competition. The hall was quite full and after the works had been played, Prof Abrahams gave his judgement on the pieces. For my work he said the following:

"When I first saw the name of Bach at the head of this piece, I thought it was unreasonable of an expert composer like J S Bach to enter a student competition. However, after hearing only two or three bars it became abundantly clear that this was not the work of Bach! I found the left hand uninteresting and the right hand equally so. In fact, you might say it is a very balanced work!"

I can't remember any more of what he said (my mind went blank!) except I think he might have said that there occasionally some interesting bits. I was mortified and after the judgement all the works were played again. Frank played J.L. Bach again and rose with his arm outstretched towards the composer for me to acknowledge the (weak) applause. I decided I was not going to get up and everyone thought Frank had written it! If you really want to hear the work, it is in Example 1 on the Website!

The following year I submitted a much more extended piano work, but the competition did not proceed that year because of lack of entries. However, a student friend of mine was studying for a music degree, and he said to me (shortly after I had submitted the work) that Prof Abrahams had come into the common room and played a piano work. He told the people in the common room that it was a work submitted for the composer's competition. He said there were faults, of course, but that he was quite impressed with the work. He also mentioned my name. However, although he returned the work there were no comments on it!

Towards the end of my time at Strathclyde, I suddenly realised that music might play an important role in HCI design, and I began to experiment with the idea. However, before I really got involved with music, I moved to Loughborough University.

Chapter 23
Computing and Music

In 1990 began to wonder if music really could be used in computers for the communication of information. What is not in doubt is that sound is a very important channel of communication for human beings generally (it is thought to have preceded vision in our evolution) yet, apart from the use of verbal communication, computers have tended to neglect other audio forms. Music, for example, is a very rich form of communication and is capable of transmitting emotion. Music can be uplifting, sad, exciting, and depressing. Audio communication also has the advantage that a user need not be attending to the computer when the communication takes place. In addition, the visual channel is now becoming very crowded, and audio offers some relief from this situation.

Computers had already been used to create sound output by 1990 mainly for alarms and for sight-challenged users. In 1993 I wondered if the use could be extended. Could instrumental sounds be used to explain what was happening in programs or shed new light on program execution (i.e. assist debugging a program)? The MIDI interface had recently become available and could easily be installed on a computer to enable instrumental sounds to be added into programs. MIDI is easy to use. Calls can be added to a program to allow output of all the instruments of the orchestra.

People have often claimed that there is a relationship between Mathematics and Music. Indeed, in the earliest times music and science were regarded as synonymous. Pythagoras considered there to be a "harmony of the spheres" with the planets and stars moved according to mathematical equations, which corresponded to musical notes and thus produced a symphony. Both Mathematics and Music are concerned with patterns and their manipulation, so it is perhaps not surprising that they are in harmony with each other. There is also likely to be connection between Computing and Music as well since

Computer Science has its roots in Mathematics. It is not known if Alan Turing was very interested in Music and his favourite tune was apparently Molly Malone, but there used to be a popular group called the *Turing Machine.*

But how could music be exploited in human computer communication? What properties of music could we use to transmit information? First, we have the notes themselves. In Western music, the standard scale is CDEFGABC (i.e. the white notes on the piano), and of course there are the black notes as well (Listen to Example 2 (White Note Scale) and Example 3 (Chromatic Scale) on the website for the two scales). There are other scales for example, the Pentatonic scale (used in Japanese and Chinese music) uses the notes CDEGA—Listen to Example 4 Pentatonic Scale). Indian and Arabian music has some differences from Western music. However, there are many similarities.

The human brain seems to like simple frequencies. A musical note is based on a single frequency produced by the vibration of a string (violin, piano etc.) or an air column (horn, bassoon, flute etc.). The different notes on the piano are all simple frequency ratios. An octave on the piano (i.e. C to the C above) is simply a doubling of the frequency and most other notes are simple fractions of the Octave. G is 1/3, E is 1/5 etc. so it appears human beings like simple vibrations.

Another major aspect of music is Rhythm. Most music has a regular beat (i.e. a march—two beats in a bar, or a waltz—three beats in a bar) but there can be different rhythms for example 5, 7 or 9 in a bar.

A further important property is Timbre. Different instruments play the same notes as the piano, but the sound is different (e.g., Flute, Clarinet, Violin, Drum). They are the same frequency but have different harmonics in them. Timbre is a useful property. It means that the human being can hear several musical notes simultaneously and distinguish between them.

A final property is that music is 3-Dimensional. So different instruments can be placed in different points in the sound space.

But what about people? Do they all hear the same notes? Do certain combinations of notes appeal more than others. The fact that we have a "Top of the Pops", where most people agree that a particular tune is better than another, would indicate that people do tend to hear the same thing. Also, some tunes are universally liked and appreciated by most people, so there does seem to be a common understanding.

One important point is that most people do not have perfect pitch. In other words, they cannot tell which note is being played in isolation. On a normal piano, Middle C is 256 cycles per second, but if it had a frequency of 270 or 280 cycles per second most people could not tell the difference. This is why orchestras have to tune up at the beginning of a concert so that all players are playing in tune.

There are many people who have claimed that there are too many difficulties in using music to communicate using a computer. For example, the appreciation of music is different in different people, and we know it varies across cultures. My view was that although there are differences between music systems, the similarities are usually closer. Also, most people are frightened of singing out loud, and they often claim that they are tone-deaf, or that they cannot sing. This is partly because people are embarrassed at singing in public. In fact, very few people are completely tone-deaf.

To see if it was possible to communicate information via music, I decided to take a popular computer algorithm and set its operation to music! The Bubble Sort algorithm was chosen because it is relatively simple and well-known. In the Bubble Sort Algorithm, a set of numerical values are sorted either into ascending or descending order. So, for example the numbers 8395642 would be rearranged in ascending order by the Bubble Sort as 2345689.

The algorithm does a pass through the numbers and swaps any numbers in the wrong order. When it has completed the pass, it does it again and continues until no swaps are required. Then the list has been sorted. It is called a Bubble Sort because the high number swaps progressively "bubble" up to the top of the list.

Let me explain the bubble sort algorithm to you.

Assume the original list of numbers is 5 8 3 2 9 6 and we want to order them in ascending order.

At the first pass of the algorithm, 5 and 8 would considered first and they would be unchanged because they are in the right order. So, we move one up the list to 8 and 3. These are in the wrong order so 8 and 3 are reversed.

The list is now 5 3 8 2 9 6.

Next 8 and 2 are considered. They are also in the wrong order and so are reversed.

The list is now 5 3 2 8 9 6.

Then 8 and 9 are in the right order so are not changed.

Finally, 9 and 6 will be reversed.

So, the list after the first pass is 5 3 2 8 6 9.

We then do a second pass.

5 and 3 are reversed. The list is 3 5 2 8 6 9.

Then 5 and 2 are reversed. The list is 3 2 5 8 6 9.

5 and 8 are in the right order giving a list of 3 2 5 8 6 9.

8 and 6 will be reversed giving a list of 3 2 5 6 8 9.

but 8 and 9 will be unchanged.

So, after the second pass the list is 3 2 5 6 8 9.

In the next pass the 3 will be exchanged with 2. After that there will be no changes because the list is in ascending order! i.e., 2 3 5 6 8 9.

It is called a bubble sort because the high numbers slowly bubble up to the high end and low numbers sink down to the low end!

I wrote a program in the PASCAL computer language to carry out the Bubble Sort and added calls, at the appropriate places, to produce sounds using MIDI calls. The speed of the program had to be slowed down appreciably otherwise the music moves too quickly to be heard properly.

The first problem is how to represent the list of numbers. The most obvious mapping was to use rising notes to represent the numbers to be sorted. The program only used a simple list of say twenty numbers, so I represented the numbers by notes on the piano staring with Middle C =1, D=2, E=3 etc. up to 3 octaves above Middle C (I also had an alternative mapping—Middle C=1, C# =2. D = 3, E#=4 etc. i.e., using both the white and black notes).

Before each complete pass, I played the current list, which if unsorted would sound musically jagged. If sorted it would sound as a smooth run up the scale. The notes were played on a Celeste. Both scales worked well. As the list is being traversed the listener needs to know which element is being tested and if a swap takes place. Progress up the list was therefore marked by the harp and if a swap took place the listener heard a set of trumpets playing a chord (with a twiddle).

Stereophony was also used to help users distinguish the sounds, the harp was on the right, the celeste on the left and the trumpets in the middle. Since Program execution is normally far too fast the program had to be slowed down by inserting pauses. At the end, when the list has been sorted, a musical cadence was used—like a musical Amen. On the Website you can hear the full White Notes Bubble sort. (Example 5) and the Black and White Note Bubble Sort (Example 6).

The musical Bubble Sort went really well, and everyone could follow it. When I played the example at conferences it always received an enthusiastic clap at the Amen! So, music really can be used to communicate information.

Composing and playing music can be a hazardous business. In October 1991, an incident happened related to my musical abilities. Peter Mowforth of the Turing Institute had fallen in love with a Programmer in the Turing Institute called Gill whom I knew well. The wedding was arranged for October 6th at Ross Priory (by Loch Lomond) and Peter asked me to play the organ at the ceremony. I never like playing in public and there was not a proper organ at Ross Priory, so I suggested I bring up my keyboard and computer. I could write the required music using software called Sibelius: Sibelius had a realistic organ sound. I could record all the music on the computer and play it at the press of a button!

Provided we hid the Keyboard behind the piano at Ross Priory everyone would think I was playing it!

By Email I asked Peter what music they would like. He replied, "Could you play *Where Sheep May Safely Graze* as the guests assemble. During the ceremony the two hymns will be "The Lord is my Shepherd" (*Crimond*) and the hymn "*Oh Perfect Love*". As the couple retire, please play *Mendelsohn's wedding march*.

However, there are two versions of "*Oh Perfect Love*" and I asked Peter which one did he want? He chose one of the versions. I then programmed the computer and left for Glasgow with my wife Mary.

On the wedding day, we set up in the main room at Ross Priory with me conveniently hidden behind the piano. I pressed the button on my lap-top and started the grazing sheep tune and it seemed to go fine until a bagpiper suddenly appeared outside and completely drowned the music. Peter had forgotten to tell me! However, it didn't really matter, and the guests assembled. The vicar was an old Church of Scotland vicar who was hard of hearing.

Things went reasonably well as I pressed the appropriate buttons and pretended to play the hymns. We got through Crimond OK, except I hadn't left enough time between each verse, so the congregation were rather breathless at the end! The vicar then said, "We will sing the devotions—organist!" Well, Peter hadn't told me about this, so I assumed that we should sing "Oh Perfect Love". I pressed the button, and no-one seemed to be singing so in desperation I sang out at the top of my voice. Not only was it not the devotions, it was also the wrong tune!

The vicar looked incredulously at me, but I pressed on with hardly anyone singing. We finally finished the three verses of "Oh Perfect Love" and the vicar then said, "We will now sing Oh Perfect Love!"

I replied, "We have just sung it."

"No, you haven't," said the vicar.

"Oh yes, we have," said the congregation!

The vicar then said, "That was the wrong tune. Play the proper one, organist!" I didn't have a button to press and so I said I couldn't do it. "We'll sing it without you then," replied the vicar and led the congregation in three verses whilst I cowered behind the piano.

At this point, Peter realised what had happened and burst out laughing followed by the rest of the congregation. I recovered enough to press the right button to start the wedding march and the bride and groom walked down the aisle. The guests had found the incident very amusing as had the bride and groom, so no damage was done. Later at the dinner, the vicar apologised for his poor hearing, but I said it didn't matter.

At the time that I was tentatively exploring music and computing, and others were also starting out on the same quest. Some work had been done on creating a Word Processor for blind users (Soundtrack, Edwards). The system used a combination of speech and square waves. Moving up and down the mouse caused the musical note to rise in pitch or fall. A graphical program had also been created called Mercator (Mynatt) for blind users and there had also been work on what were called "auditory icons" or *Earcons* by Blattner. These are a short sequence of notes in a sequence which is easily memorable. "Ta Daa" is a simple example.

After the success of the Bubble Sort, I wondered if we could use music in a more quantitative manner. People can usually tell if notes are rising or falling, but can they detect actual note changes? For example, if a C is played and then an F, could listeners tell that if C=1 then F=4? Choristers can do this (i.e. C to F (a fourth) or C to G (Fifth) but could the average person? Try—Example 7 where I have given a set of two notes in succession. Can you tell the difference? i.e. if the note C is followed by F what interval do you think it is? (it is actually 4). The answers are on the website.

To test whether music could be used quantitatively, I set up an experiment where we played a set of tone intervals (i.e. CE, CF, CA etc.) and asked subjects to say what that interval was. We used intervals rather than absolute notes because very few people have perfect pitch (i.e. can identify a note exactly). The results are as follows.

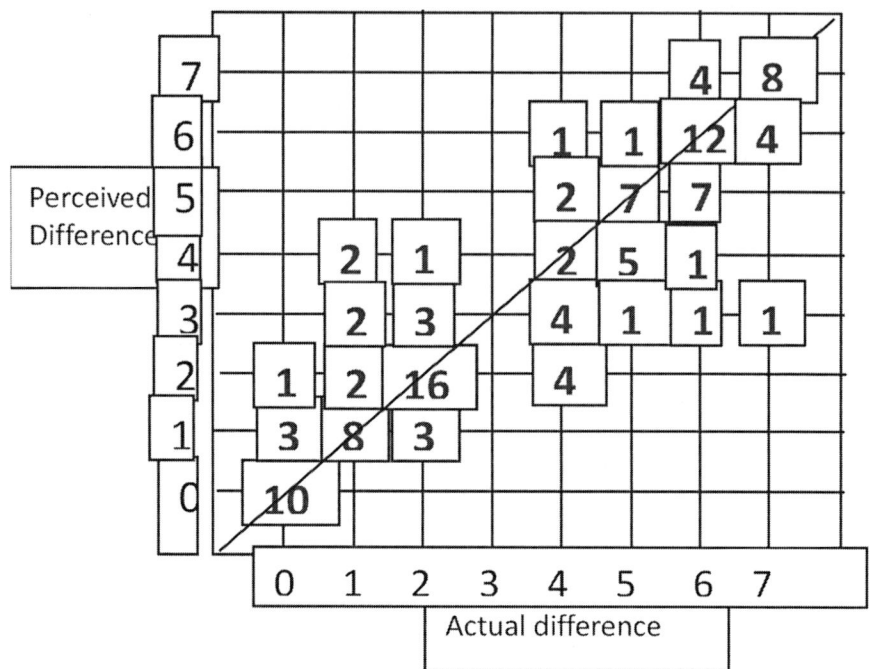

In the experiment, 62% of subjects were 100% correct, 28% achieved a difference within 1 note and 9% a difference within 2 notes in the intervals. When the second note was far away from C (i.e. far from each end) subjects sometimes were 1 or 2 notes out. Subjects were less accurate for intervals of 3 and 4 than at each end. However, musicians achieved an accuracy of 93%. So, although the tone difference cannot be used for accurately determining the numerical value of a note, it is still useful for detecting shifts. In this experiment there was an error—no 3-note interval was presented.

This was why the Bubble Sort Algorithm worked so well. Absolute accuracy was not needed, a subject only needed to know that the pitch was rising or falling. We also asked users to sketch the shape of a musical sequence. Generally, the sketches were acceptable, but if a recognisable tune was played, the accuracy improved.

The Bubble Sort was well received at various conferences and in 1994 I had approaches from two prospective PhD students who wanted to study for a PhD using Computers to communicate via Music.

The first student—Dimitrios Rigas—was interested in communicating graphical diagrams using Music and Computing. The idea was to create a program which would, using music alone, communicate the content of a diagram (i.e. circles, rectangles, lines, ellipses, etc.), to blind users. Dimitrios was very keen but was not a musical expert (although he appreciated and enjoyed music). I therefore supplied the musical knowledge.

Much of the use of music in computers had been limited to simple tones (i.e. like the notes of a tuning fork). We considered that much more could be done using rich musical stimuli from real musical instruments.

Once Dimitrios' program was fully working, a set of subjects performed the experiment. All users were visually impaired in some way, but all had some visual experience. They had no exceptional musical ability, however, and to make the experimental conditions the same, we blindfolded them all. The subjects were then asked to listen to a diagram communicated solely using music (which contained circles, lines, and rectangles) and were then asked to draw the diagram from listening to the music alone. The only problem we had to solve was how the user drew the result (whilst blindfolded!). We therefore gave them a sheet of paper with the axes on it and put raised 64x64 grid lines on the papers so that they could locate the positions by touch. The results were quite good. All diagrams were correctly perceived, generally in the right location but not accurately placed.

Dimitrios began by trying to transmit the position of a point (x, y) on a graph, and did this by simply going up the musical scale, first for the y axis, and then for the x axis, using two different instruments. The scales had 1—40 on each axis. A piano was used for the x-axis and a harp for the y-axis. So, if the coordinate was (8,4) the piano would play C, D, E…up to C, and the harp would play C, D, E, F (Listen to Example 8).

The first experiment asked the listeners to locate the point being transmitted by music. When Dimitrios presented the results to me, I was initially astonished at how well the users had done (blindfolded) and couldn't believe it. I therefore immediately went to the laboratory. Dimitrios blindfolded me and I did the experiment myself and confirmed his findings. Dimitrios then traced out the various figures (rectangle, circle line) by simply playing, in sequence the coordinates of the figure. So, a rectangle would start at the bottom left-hand corner and the music would trace the shape by playing the succeeding (x, y) coordinates round the perimeter. Although this took a few seconds, it worked well. (Example 9 is (9, 6) using piano and harp).

A line was simply the progressive (x, y) coordinates (rising or falling) depending on the slope (Listen to Example 10). Dimitrios called the program AUDIOGRAPH. The complete diagram was presented by successively scanning each object in the diagram in turn from top-left to bottom right. Here is an example, and the user response.

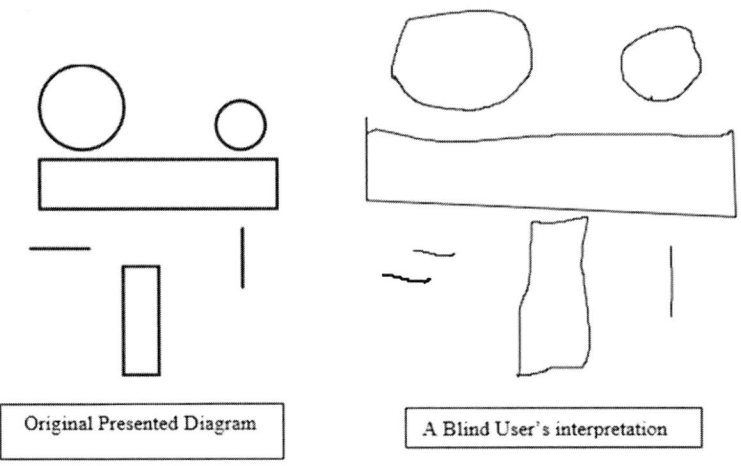

Original Presented Diagram | A Blind User's interpretation

Although the results were encouraging, we felt that if the subjects were given some clue as to what they were perceiving, that would enhance recognition. For example, we might try to communicate a letter of the alphabet, a road vehicle, or a map of the University Campus. We therefore presented four diagrams—a road vehicle, the letter E, the number 3, and a histogram. One set of users were given the clue as to what the figure represented (i.e. a car), and the other group received no indication of what the figures were. The group who had received the semantic information performed much better than the other group. When the car shape was presented with semantic information, recognition rose from 0% to 100%.

The only disadvantage of the system was that the presentation of a diagram took much longer than one would demand in a commercially available tool. However, it did show the importance of context. Dimitrios obtained his PhD on this work and has since done extensive work in the area.

The second student who approached me to study for a PhD (part time) in 1994 using music in computing interactions was Paul Vickers (a Lecturer at Liverpool John Moore's University). He had seen my paper on the Bubble Sort.

He wondered if Computer Program Debugging might be an interesting topic. I thought it was a great idea and Paul and I worked closely together over the next 5 years, and he obtained his PhD in 1999. Paul had quite a good working knowledge of Musical Theory.

We both realised that some of the earlier failures which attempted to utilise music failed because the designers were not musical. We reasoned that any researcher trying to communicate information by music must understand how music is constructed so that appropriate mappings can be made between the artefacts (the characteristics of the object being auralised) and similar musical structures. For example, most people can recognise a major or minor chord (though they may not understand what they are). The major chord is bright, and the minor chord is sad, so success or failure, say, in an IF statement, could be signalled by a major, or a minor chord and users would intuitively recognise this (Listen to Example 11).

To aid debugging we were attempting to represent the various structures used in Programming. For those not familiar with computer programming the PASCAL computer language uses what are called **CONSTRUCTS**. There are two basic types—**SELECTION** and **ITERATION**. The **SELECTION** statements do a particular action depending upon the value of something. The two main types are **IF THEN ELSE** and **CASE** constructs.

Here is an example of an **IF THEN ELSE** construct:
IF Today is [Weekday] **THEN**
WriteLn ('Must work harder')
ELSE
WriteLn ('Take a day off.');
END:

This statement checks if a condition is true (i.e. is Today a weekday) and if this is true it carries out the statement after **THEN**, if not it carries out the statement after **ELSE**. If the value of Today is Monday …Friday, then the program writes "Must work Harder". If Today has any other value it writes "Take a day off".

Here is an example of a **CASE** statement.
CASE (grade) of
'A': writeln ('Excellent!')
'B', 'C': writeln(Reasonable")

'D': writeln("You Failed")
END;

So, if grade has the value "A" the output is "Excellent". For "B" and "C" the output is "Reasonable" and for grade = "D" the output is "Failed".

The other main type of construct are **ITERATION** constructs. These carry out a series of actions until some condition is achieved. Two examples are **WHILE ()** and **REPEAT UNTIL**.

Here is a **WHILE** statement. This will write out the value of *Number* for values 2, 3, 4, and 5. And will stop at number=6. It carries out the statements between the **BEGIN** and **END** until number=6.

Number := 2;
WHILE *Number* < 6 **DO**
BEGIN
writeln (*Number*);
Number := *Number* + 1
end;

Here is a **REPEAT UNTIL** loop
REPEAT
S1;
S2;
Sn;
UNTIL condition;

This statement carries out S1, S2, ….Sn, until the condition is met.

The Musical Cues which were applied to the PASCAL structures were designed as follows:

In normal speaking, for **SELECTION**, we upwardly inflect our voices when we ask a question, and the answer usually drops to a lower level. So, for **SELECTION** constructs we adopted a rising set of notes at the start and a descending set on exit.

For an **ITERATION** construct a background droning sound was used to show that continuous action within a construct was being carried out. Then it changed to a "Success" sound (i.e., a pleasant sound) on completion.

On the Website:

Example 12 illustrates a successful IF statement (listen for the major chord) and An Example 13 an unsuccessful IF statement (Listen for the minor chord).

Example 14 illustrates a CASE statement.

Full details of the Auralisations can be found in the research paper (Vickers and Alty, Interacting with Computers, Vol 14, (2002), 457-485). In all, 60 constructs were used in the experiment. Half were iterations and half were selections.

The first part of the experiment checked if students could recognise the different constructs. On the whole students were able to visualise program structure using only the musical sound as the cue. Often, even if a mistake was made the answer was still in the right construct class. It was also interesting that no significant differences were found across subjects with varying levels of musical knowledge.

The main objective of the experiment was to see what contribution the musical cues had on the task of finding errors (called debugging) in computer programs for inexperienced programmers. One advantage of the aural approach over normal debugging techniques (where you have to examine the program list on paper separately) is that it is immediate and can be generated as the program is running.

We limited the examples, which were presented to students, to PACAL constructs and errors which directly or indirectly manifested themselves in errors in program flow. Could the subjects locate the position of a program error (bug) within a given program text just by listening to the music? We also wanted to know:

1. Could novice programmers locate more bugs using the musical sounds compared with using the traditional paper listings?
2. Did they locate bugs quicker with musical cues than without them?
3. Does their musical background (i.e., their familiarity with music) affect their ability to use the musical sounds as cues?

It was quite a complex experiment and too detailed to be reported here but an interested reader should read a rather aptly entitled Journal Paper which we wrote with the title "When Bugs Sing"! (I thought this was a really good title!).

Musical Cues did assist with bug location. We did not see any difference in the times taken, so the Musical Cues did not alter the time to complete the debugging exercise. However, it did considerably increase their workload. We also found that past musical experience did affect their ability to use the Musical Cues.

Finally, we provided some guidelines for future designers interested in using Musical Cues in their work.

1. Use tonality. Do not use disconnected sounds. Western-culture music is widely understood across the globe and the standard diatonic scale is well-known. Major and Minor keys are generally well-understood.
2. Use a hierarchical structure. Any related collection of objects to be communicated should map into similarly organised musical constructs.
3. Metre and Rhythm are more easily retained than Harmonic Structure.
4. Percussive devices (i.e., drums) should be used to add emphasis. Drones (Long held notes) are important to maintain continuity over constructs.

Paul obtained his PhD in 1999 and was highly commended by the External Examiner. Paul was unusual in that he did his PhD externally whilst holding down a full-time job.

In all I think I supervised 16 students studying for a PhD whilst at Loughborough. All were very competent and produced successful theses. I was always impressed with students who completed a PHD externally. Another student who did an external PhD with me was Raj Curwen. Raj was a very bright student. Raj's thesis was concerned with improving search systems. He argued that, because of the wealth of information on the internet – picture, music files, information pages and documents had become so large, the search process needed to be able to incorporate all these different data sources in a single search and to be successful the context of the held data and its relevance to the search must be understood. Raj therefore developed a search engine based upon the use of context. He did this by taking account of the actions which users carried out whilst working with their files and he coined the term "machine acts" to describe such acts. Such acts are stored and indexed with the files acted upon.

He developed a prototype system which was able to sense context as the user interacts with the files and stores this as machine acts. He carried out a series of experiments to test the idea. The results were very encouraging, and Raj has now moved on and formed his own company. He is developing a new computer language which should enable students to code much more quickly and efficiently. In all I think I supervised 16 students studying for a PhD whilst at Loughborough. All were very competent and produced successful theses.

Chapter 24
The Digital Audio Broadcasting (DAB) Project

This project is different than would be expected from an Academic Research Group. Our expertise in Human Computer Interaction and in Telecommunications was fully utilised, but the project was very much a development project rather than research. However, it had a significant impact on the growth of Digital Audio Broadcasting and on DAB Receiver Design, both in the UK and the rest of Europe.

So, what is Digital Audio Broadcasting (DAB)? Radio broadcasting was introduced in the 1920s and the first television broadcasts began in the UK in 1936. Over the next 60 years, both these used analogue signals (a continuous electrical wave, rising and falling). With traditional television and radio, the signal is unidirectional (i.e. information only goes one way, from the transmitters to the receivers). The listener or viewer cannot interact in any way with the broadcast. (i.e. send information back directly back to the broadcaster). They can interact using the telephone, but this is very cumbersome and only a limited number of phone-ins can be supported.

During the early nineties there were plans to make British television digital, and implementers were looking at making radio transmission digital as well.

In 1996, the European International Standards Institute had introduced the first commercial completely digital system for terrestrial audio broadcasting. It was introduced because digital broadcasting offers many advantages over analogue broadcasting—better management of the signal, lower cost for transmission, cheaper receivers, better mobile reception, and higher sound, speech, video, and data quality.

Importantly, digital transmission ignores the type of data being transmitted (audio, video, and text for example) and this means that the services offered remain independent of the transmission system. Also, several digital services can share the same transmission, again making it more efficient, but what was really new in DAB for broadcasters was the possibility of transmitting more information per second and more accurately.

Because the information transmitted is much greater and can include pictures or text, DAB offers a sort of "half-way house" between traditional radio and television. For example, radio signals could be enhanced with additional pictures or music. 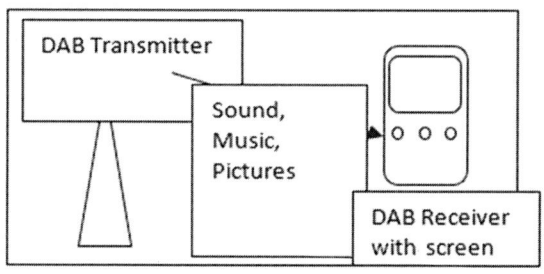 However, this would require the receiver to display pictures as well as the sound. So, if a presenter was talking about say, an historic mansion, pictures of the various rooms could be transmitted alongside the commentary. Of course, the receiver would need a screen to display the picture, but with the falling costs of data storage, this could be provided cheaply at the receiver end. In addition, the receiver could have some microtechnology in it so that it could further process the picture.

The receiver, therefore, is no longer just a loudspeaker. With DAB the receiver can now have a screen on which additional information can be displayed such as text, pictures, graphics and video clips, in addition to sound.

However, DAB offers another important advantage. In more recent times, broadcasters have attempted to get viewers involved in TV programmes. For example, a programme could ask listeners to vote on the topics being presented but listeners could only do this via the phone (or email). This is very limiting because there are not enough telephone lines for incoming calls. In Chapter 13 I told you about the famous TV programme on Sunday pub opening in Wales, when all the telephone lines were jammed because thousands of people were trying to object to one of the TV guests. A digital return channel could easily have coped with comments from thousands of people. The presenter could have asked people to vote for or against the guest and a computer could, in seconds, have displayed the result. The introduction of an input channel, therefore, could lead to really exciting radio. Although the DAB download channel (i.e., the broadcast) will need to carry a high amount of data, the upload channel could be quite modest in capacity.

There are lots of examples of where a broadcast in conjunction with a return channel could be useful. One example might be the production of a newspaper. Traditionally, high volume printing presses produce thousands of copies which then require 100's of delivery vans to get the papers out to the newsagents very quickly across the country over-night. Using DAB, the papers need no longer be physically delivered. Instead, newsagents would install high speed printers and the papers could be printed on demand as readers came into the shop downloaded via DAB. The upload facility would allow the newsagent to order more copies of the papers and perhaps different editions. A DAB system could also offer a city information service (for example, offering current traffic condition to approaching car drivers), a data delivery system on a building site where there is no hard-wired network, and could be used in an educational setting (such as computer aided learning).

My research group became interested in DAB in 1995 largely because one of my Research Assistants, Iain Duncumb (who had originally been a manager in the telecommunications business and was very talented) suggested that DAB was an intriguing development which was being ignored. Iain was an interesting person. He was comfortably off and came to me because he was interested in taking part in challenging research work at the University. He loved the technical challenge but was not interested in submitting for a PhD.

I had never heard of DAB until Iain told me about it in 1995. At that time only a few expensive car radios could receive DAB and they had a small screen telling the listener what text was being broadcast alongside the sound. However, because adding a screen to a receiver is necessarily more expensive, this seemed to have discouraged manufacturers at the time from offering portable receivers with a screen. He also pointed out that there had been no progress on developing the return channel (or up-channel).

I asked Ian to check it out. Iain worked hard and eventually brought together an interesting consortium of industrialists to try to get DAB (including the up channel) fully adopted. Besides our research group, the main industrial partner was the German company Robert Bosch GmbH, but also included were Teracom AB from Sweden, Deutche Telekom AG, WestDeutscher Rundfunk (WDR) from Germany, Télédiffusion de France and the Universities of Nottingham, Napier University and UCL London. Radiotelevisione Italiana and Ericsson Radio Systems AB also joined the project in the second year.

In September 1995, the consortium was awarded a large contract from the EU Advanced Telecommunications Directorate. The project was called MEMO (Multimedia for Mobiles). It had "Mobiles" in the title because transmitting DAB to mobile receivers seemed one way to really exploit the potential of DAB. As part of the consortium my Loughborough University research group was awarded a £350,000 contract mainly to work on Human Computer Interaction aspects of the project so that users could easily interact with the broadcasts. The basic goal of the project was to specify, and to demonstrate a communications system to convey both short and long messages in the Up channel in a flexible way using DAB. The proposed system would use high speed downloading to mobiles and the users could then interact with the broadcast with an Up channel using GSM (Global System for Mobile Communications). By 2014, GSM had over 90% market share, operating in over 219 countries and territories.

A huge set of possible application areas was envisaged using the Up-channel, and this was why so many national hosts and companies were willing to join the consortium. Application areas included—downloading a newspaper, construction plans for building sites, interactive games, and traffic and travel information.

The first task was to get the upload channel working. Bosch DAB-Receivers were used in the demonstrators. A 320Kbs DAB service was demonstrated at Rennes in July 1997 and Eriksson demonstrated a DAB Down-link and a GSM Up-link at Gothenberg in November 1997. In November 1997, an article appeared in the times newspaper entitled "Radio with Images" and it included an interview with our project leader.

Two applications were demonstrated by Teracom in Berlin in August-September 1997 at the International Funkausstellung (a very important forum which had 436,000 visitors, of which 150,000 were professionals). The first demonstrator was a Traffic and Travel Application. Traffic announcements, concerned with the location of the receiving vehicle, were received by DAB and plotted on a map on the receiver in the car. The application also provided the location of petrol station and hotels, and the user could book a room at the hotel using the uplink.

The second application, the newspaper application, was run on a small mobile PC and demonstrated a possible look-and-feel computer-based news service. This was the only application at the conference demonstrating the combination of DAB and Mobile Phones and it generated a huge amount of interest from government minsters, journalists and professionals as well as the general public. West DeucherRundfunk (WDR) presented content with actual news, traffic information with pictures, maps, weather information, and timetables for the arrival and departure of aircraft services. The Loughborough research group was fully involved in all the interface design issues during these trials.

The project had a major review in January 1998. The technical audit was a great success. They asked us to trial at least 15 simultaneous users at the final trial. At the end of the project, the Project Office congratulated us on successful completion of the project. We and our partner had proved the usefulness of the DAB concept. However, one problem that remained was the lack of take-up by the DAB Receiver Manufacturers.

As I mentioned earlier, implementation of DAB in the UK (and in Europe) was progressing far too slowly. Iain showed me an article from the Sunday Times which discussed the problem. Apart from two expensive radio receivers, made by Grundig and Pioneer (both costing £700), there were no other receivers on the market. In particular, there were no *portable* receivers, and no UK or European manufacturer was developing one. Because DAB was 100% European and the Americans were reluctant to endorse it, it had meant that the Japanese manufacturers (such as Sony, Sanyo and Aiwa) had also ignored the development.

Iain argued that a reasonably priced receiver, with a really simple but effective Human Computer Interface would appeal to all sectors of the community including the elderly and the disabled. It ought to be possible using the digital approach to make a portable receiver which would be easier to use that the analogue equivalents. Furthermore, we could utilise our expertise in Human Computer Interaction to design a really simple-to-use receiver. Iain also suggested that a DAB Up-channel internet audio link should be included so that a listener could interact with specialist services over DAB.

In August 1998, therefore, we got together a consortium consisting of Roberts Radio (the one remaining receiver manufacturer in the UK) and World Radio Network Ltd (WRN) and submitted a bid (appropriately called PORTADAB) to develop a portable DAB receiver to the Department of Trade and Industry (DTI). The consortium would develop an easy to use, relatively cheap DAB receiver which would have a low-capacity internet Up-channel. One hundred sets would initially be produced. We heard in the summer that the bid had been accepted. The total grant awarded was £277,000 of which my research group at Loughborough received £170,000. The project was led by Simon Spanswick of WRN.

By May 1999, the physical design of the receiver was well advanced. It would have a rechargeable battery to give about 5 hours of operation before recharge. DAB receivers are very different from standard receivers and a new interface design was required so that the users would have no difficulty in using the receivers. This was Loughborough University's responsibility. Two rotary knobs, a combined on/off and volume control, and a tuning knob were incorporated in the design and there were also four pre-set buttons so users could easily select their preferred channels. The display would be backlit and would be 4in x 3in and cost about £80.

At the meeting we presented the first designs for the Human Computer Interface and provided a number of options.

1. The radio will be switched on/off by pressing the volume control knob and a separate knob would serve as a tuner.
2. The list of available services would be explored by rotating the tuning knob. Available ensembles and services will be selected after a 5 second press. If the tuning knob is not touched it will automatically tune to the last service chosen in 10 seconds.
3. Four pre-set buttons will be provided to enable the listener to select one of four pre-set channels and the buttons will be labelled with an 8-character user chosen label.
4. The current time would be displayed in hours and minutes. The battery strength would be displayed when the radio was on. One line of the display would be used to assist the user.

5. There will be an engineering display (not available to the normal user) which would be selected by pressing a combination of buttons. It would display frequency, band and mode, transmitter IDs, field strengths,

channel ID's, error estimates and the strength of the audio signal. Return to normal service will be achieved by pressing the on/off button.

A full prototype of the receiver prototype (right above) was demonstrated at the Project Meeting on 4 November 1999, and all participants were impressed with the progress made.

Meanwhile, there was progress in the establishment of DAB services. In November 1999, the Sunday Times declared that two new digital-only networks were being announced that week—Core and Planet Rock. In addition, Capital Global was going to make its debut in London in 2000 and the Radio Authority announced that they are going to award regional DAB licences in 2001. The Sunday Times also announced the important development of our new receiver and DAB was beginning to be taken up!

We made a few improvements to the receiver. Beeps were made when buttons were pressed and the characters on the screen were also enlarged to assist partially sighted users. A graphic was added to show the volume level. An option was also added to enable a user to have a different configuration when in a different geographic area. At the same time a new engineering page was added to allow the Human Computer Interface configuration to be altered by engineers.

In February 2000, 10 models were ready and the BBC were expected to take the other 90. Trials had indicated some reception problems in different parts of a listener's house. This was addressed successfully. I understand that Prince Charles and Harold Wilson were given one of our radios.

In March 2000, a prototype of the receiver was delivered to the BBC for evaluation. The top line comment was that "the receiver delivers the ease of use in a compelling way". This was quite an accolade for the Loughborough design! It had also been shown at the World DAB meeting and the BBC commented that the users found our new radio very easy to use. They had some criticism of the speakers (too small) and they worried whether the radio was robust enough. The Gramophone Magazine reviewed the receiver in August 2000 and commented that the initial £800 price tag should drop to £500 by the time volume production begins. Roberts also announced that they were planning a cut down version to sell for £200. The general conclusion of users was that the DAB receiver was much easier to tune and to set using the pre-sets.

Shortly afterwards, Roberts produced the much cheaper version of the receiver. This successfully entered the market and began to sell well, so our work was complete. Shortly afterwards many other receivers began to enter the market.

We were delighted that our project had met its overall objective of promoting DAB. Interestingly, the original DAB receiver, which our team at Loughborough designed along with Roberts and RealWorldNetworks has become a collector's item! In fact, there is one of our receivers on display in the British Science Museum!

Although we were delighted with the result, we argued that software was needed to enable program developers to easily create new applications. We therefore proposed a new project to the DTI.

The new project was set up in partnership with Nucleus Digital, Switchdigital and TTPCom. These were good partners to work with. Nucleus Digital had been formed 1999 by experienced TV producers to produce broadcast programmes and explore new technologies and Switchdigital had gone on air June 2000 with six digital stations including Heart, Jazz FM and BBC London Live. Later Unique Interactive also joined the project.

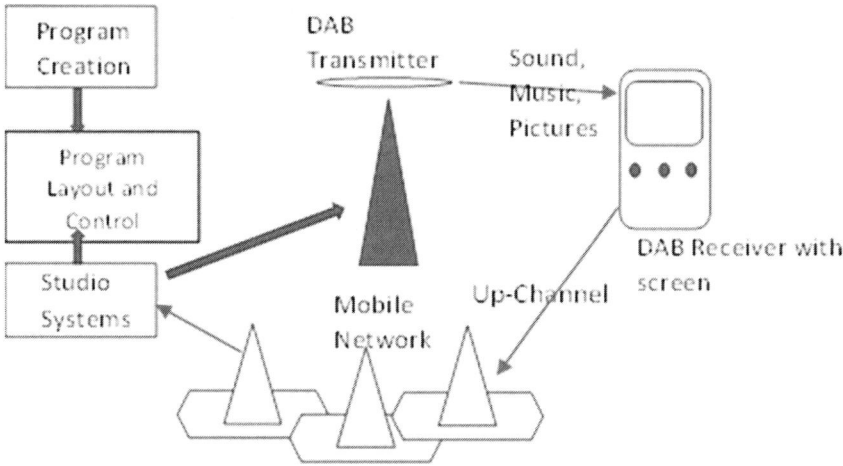

The grant proposal was submitted in July 2001, when 80% of the population of the UK were able to receive DAB and one objective of the project was to maintain the UK as a DAB leader. The grant of £350,000 was approved in August 2001.

The project developed a low-cost user-friendly system for producing a DAB program including sound editing, connecting pictures to the sound, tell-me-more options and the ability for the listener to interact with the program. Such programs were expected to make watching the radio activity an exciting choice of the future (by adding a display screen and phone to the receiver). The key idea was for multimedia and interactivity to add to the audio backbone of radio. The new approach was expected to encourage audience feedback from music programs, map diagrams, current affairs programs, popular shows and advertisements. One key feature of radio is that radio is listened to without the listener necessarily giving full attention. It was important for the project to ensure that this feature would be retained whilst offering interactivity.

In November 2001 a number of program scripts were developed for evaluation purposes—Astronomy, The Peter and Geoff Show on Virgin, a gambling application and a voting system. The project was a success. These days, anyone with a DAB receiver can now see multimedia material being presented alongside radio programs. (the BBC planned to send maps, diagrams, and photos along with the broadcasts). At the end of the project the Open University also became interested in using DAB.

Because of the success of the two above projects a bid for the establishment of a DAB Laboratory at Loughborough University, run by my research group (together with a DAB transmitter), was submitted in October 2000 to the Link program. It had industrial sponsorship of £80,000 from BT, Eddystone Radio, Nortel Networks, Roke Manor Research, BAE, Roberts Radio, and Oak FM (Loughborough local radio station). It was approved under the link program on 1 April 2001 and was supported until 1 July 2003. One of the earliest uses of the transmitter was to initiate DAB transmissions by Oak FM in Leicestershire. Interestingly the first piece of music transmitted over DAB in Loughborough was a Fanfare for Brass Instruments written by me (!), originally for the opening of an HCI'2000 Conference in Glasgow. Such is fame!

Since then, DAB transmission has become the norm. However, the more advanced applications to date have been largely ignored. If you listen to Radio 3, you can now get information on your receiver or car screen about what is being played. However, the Up-link facility (i.e. the possibility of the listener interacting with the program) has not yet been implemented, which is a shame.

Chapter 25
The Interactive Learning Project: Dyslexia Studies

In 2000, a new Post-Doctoral Research Assistant joined me called Nigel Beacham. I had known Nigel as an undergraduate at Loughborough and had lectured him in Computer Science. He had received a 2.2 Honours degree in Computer Science, and I know that he was a bit disappointed with this result, but one cause of this was that he suffered from Dyslexia. After he left Loughborough, he overcame his Dyslexia, and he later successfully obtained a PhD. The fact that he had suffered from Dyslexia became very important later in the research.

As the digital age developed, people were still taught mainly by traditional methods. A lecturer would stand up in front of a class using a blackboard and chalk to explain things. In the seventies, the blackboard was replaced by a foil projector with a large screen. Then the projector was replaced by a computer-controlled slide projector which could also project text, animated diagrams, and videos. Thus, teaching became quite sophisticated with the lecturer using voice, text, animated diagrams and video.

It therefore became possible to replace the lecturer with a computer program which could provide a similar teaching experience, and this had considerable advantages. Students could learn in their own home and at their own pace and interactive learning programs began to become popular. However, there was a drawback in the approach. The lecturer was no longer there and could not respond to individual student learning problems.

Voice, text, animated diagrams, video and even Music are called Output Media and combinations are termed Multimedia. Although it is obviously useful to use these different media to enrich the learning experience, we wondered if different combinations of media might affect learning in different ways. For example, could some combinations be more effective than others in particular situations? For example, if material was presented as diagrams accompanied by a voice-over explanation, would more learning take place than if text alone had been used? Or would an animated video result in better learning than static diagrams? In addition, our existing experience of using multimedia in power stations would be useful to the project.

Of course, effectiveness of combinations of media used might also depend to some extent on the material being communicated and on the learning characteristics of the student, since it is well-known that students learn in different ways.

One can think of good examples where different media might affect learning in different ways. If a lecturer was trying to get across how, say, a car engine worked, I think that most people would expect a carefully made video to communicate more information that a textual explanation. Sometimes, however, there are surprising results. For example, as I have already mentioned, it has been suggested that the Weather Forecast is often communicated better on radio than on Television.

We looked at what research had already been done. Mayer had used a number of different multimedia presentations to explain, for example, how a lightning storm developed, how a car's braking system worked or how a bicycle tyre pump worked. The material was presented in different media combinations and the subjects' remembering and understanding of the material, was measured in a series of tests. The results obtained gave rise to a set of design principles about multimedia design. Whilst these were useful studies, we felt that the studies were too short and did not involve enough challenging material.

We therefore decided to carry out a series of multimedia learning experiments on a much more complex domain to see how well the results would scale up. One problem with using complex domains is student motivation. If the material is complex and there is no over-riding goal to motivate the students, it is likely that students will lose interest. A learning domain was needed that students **had to learn** as part of their studies (and therefore they would be highly motivated).

A domain within a university context that is acknowledged to be inherently difficult, and yet for many students constitutes an essential skill to attain, is the domain of Statistics. Most Master and Doctoral students require statistical knowledge for analysing their experiments, and this sub-domain of statistics is reasonably compact.

At the time, in the Computer Science Department at Loughborough University, there was a M. Sc. course on Multimedia Interface Design, and an important aspect of the course was the design and evaluation of different interfaces using statistics. I had lectured for many years on this course and a major part of the course involved using Statistics to analyse experiments results, so this was an ideal course in which to test our ideas.

The statistics material is typically taught in four one-hour lectures on the course and covered basic information in Statistics. The material was presented in the four different modules (The Null Hypothesis, The Binomial Distribution, The Ranking Distribution and the Normal Distribution). If you don't understand Statistics, it is not important, but they are four different and important aspects of Statistics. Four computer-based teaching modules were therefore created lasting between 12 and 16 minutes each to teach these four subject areas.

These four modules were presented using different combinations of three media—text, diagrams, and the spoken voice. The three combinations of media chosen were:

1. Text Only—The screens consisted of text only (no voice over or diagrams).
2. Written Text + Animated Diagrams. Text interspersed with animated diagrams.
3. Spoken Text + Animated Diagrams. Text above replaced with a voice-over.

These three media combinations are typical multimedia presentation combinations used in many e-learning situations. We suspected that students would usually show overall improved learning when information was presented using the Sound + Diagrams or Text + Diagrams compared with Text alone.

The diagrams and the text were progressively built up in synchronised stages using a development tool called Macro-Media Flash. In all these presentations the students passively watched the presentation without interaction. If you would like to watch examples of the three approaches, look at the various presentations on the Web Site. Text-Only (Turn off your Volume), Text + Diagrams, and Sound + Diagrams. I couldn't find the original presentations, but these are preproduction versions (the sound quality is not so good, but you can follow them OK).

The three different media presentations were based upon identical material (for example, the written text and spoken text were identical). The diagrams were presented on the left-hand side of the screen and the text on the right and the audio presentation used the speakers. An example screen from a Text + Diagrams presentation is shown in the Figure below, giving an example of what a Hypothesis is.

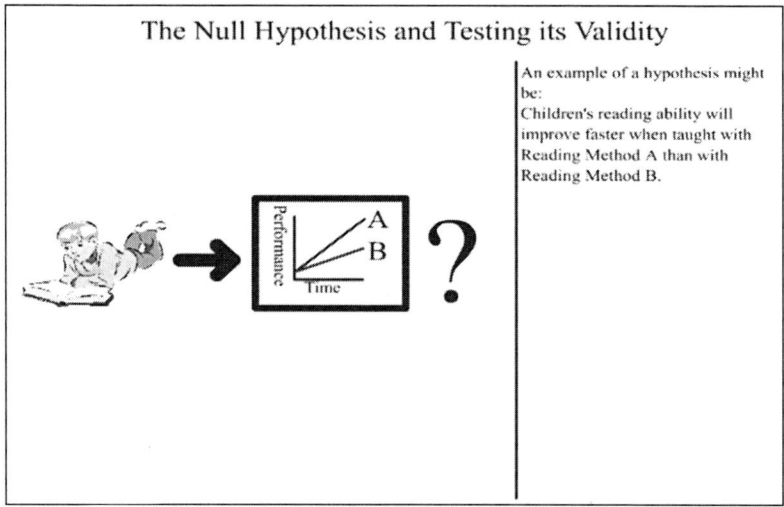

An Example Screen

Students on the M.Sc. course were all graduates and from a variety of disciplines. They were told that the four created modules would be examined at the end of the semester, but, because students would experience different styles of presentation during the experiment, there was a risk that, if presentation style really did make a difference to learning outcomes, some students might be disadvantaged in their examination depending on which presentation they used. To avoid disadvantaging students, a repeat of the presentation in a standard lecture format was given by the same tutor to all students after the conclusion of each experimental module presentation.

At this time, I was seconded to the University of Melbourne, Australia for four months as a Research Fellow with Dr Steven Howard and I went out there and worked on the presentations. Dr Howard made some really useful suggestions for the design. Nigel Beecham came out to join me for the last month.

I gave a number of business seminars in HCI in Australia and, at one of them, Nigel and I tried out the first module with the attendees (Text, Text + Sound, Text + Diagrams).

In the evening after the course, we puzzled over the results. The Text + Diagrams had done much worse than the Text-only, which didn't really make sense (it was completely contrary to existing theory). However, Nigel suddenly realised that we had made an error. We had not presented one module as Text + Diagrams, but as Text + Sound, a combination we did not intend to present. Adding identical spoken words to displayed text is not helpful and unsettles users, and theory predicts that this will be a bad combination, so the result actually supported the theory! The other demonstrations worked quite well.

On 11 September 2001, I was working my final week at the University of Melbourne. I had really enjoyed working with Steve Howard and The University of Melbourne had kindly provided me with accommodation in the University. My wife had come out for about a month, and she had just gone back to the UK. I was due to return four days later. That evening I decided to do my ironing and at 10pm I put the television on because I was so bored!

At about 10pm there was a news flash—a plane had crashed into the World Trade Centre in New York (this was early morning New York time). Pictures were rapidly put on the screen, and it was assumed this was an accident. There was an interviewer discussing the accident with the upper floors of the World Trade Centre burning in the background. Whilst the man was talking, we suddenly saw a second plane fly into the second tower of the World trade Centre! This was the start of the infamous 9/11 terrorist attack on New York. Purely by chance. I had switched on the TV at the right moment and for the next 5 hours I saw the whole incident unfold before me. It was terrifying and you simply did not know what would happen next. A third plane later crashed into the Pentagon and a fourth plane was on its way to crash on the White House. The passengers managed to overcome the terrorists on the plane, but lost control of the plane and it crashed in Pennsylvania. It was a horrific incident, and I watched the whole incident as it happened. Four days later I flew back to the UK. The security at Airports was intense and it was a worrying experience.

I went back to Australia and Steven Howard suggested that we might find interesting effects if we took into account the different ways in which students learned. He thought that the way students learn (called **Learning Style**) might be important. Students learn in different ways. Some students like to approach learning in a bottom-up fashion. They learn incremental bits and gradually build up the whole picture. Other students prefer a top-down approach. They want the high-level ideas first and the detail later. It is often said that the first lot can't see the trees for the wood, and the other lot can't see the wood for the trees!

Steve Howard suggested that we adopted the Learning Style Model proposed by Felder and Solomon. This model characterises learning style on four major axes:

1. **Sensing** versus **Intuitive** Axis

Sensing Learners prefer facts, whereas **Intuitive Learners** like to discover possibilities and relationships.

2. **Sequential** versus **Global** Axis

(Like the top-down, bottom-up distinction above).

Sequential Learners prefer to learn in incremental steps, whereas **Global Learners** prefer to grasp the general concepts first.

3. **Active** versus **Reflective** Axis

Active Learners prefer to rush in and immediately do things. **Reflective Learners** like to reflect at the beginning before rushing in.

4. **Visual** versus **Verbal** Axis

Visual Learners prefer see pictures or animations. **Verbal Learners** prefer written text or being spoken to.

From reading the above do you sense what type of learner you are?

One important reason for choosing this Learning Style approach was that it had previously been used in scientific and engineering situations. The test is also easy to administer.

The position of the learner on the four Felder Axes is determined by administering a test with 44 questions about attitudes 11 questions on each of the four axes. The test results are expressed as an odd integer (1-11) followed by the letter "a" or the letter "b" (e.g. 7a or 3b).

The "a" and "b" refer to the polar styles (i.e., 1a highly Active, or 11b Reflective) and the integer is the strength of the tendency towards that style. Thus, a 9a on the Active/Reflective axis suggests a strong tendency to an Active style, whereas a 9b would indicate a strong tendency to a Reflective learning style. A 5b would mean a mix of Active/Reflective learning. Usually, the different learning styles are spread evenly across the population as a whole. However, on the Visual/Verbal axis, there is usually a predominance of visual learners.

Two example questions from the Felder test are given below. Question 17 is concerned with Active/Reflective Learning and 20 is concerned with Sequential/Global Learning

17. When I start a homework problem, I am more likely to:
a) Start working on the solution immediately
b) Try to fully understand the problem first
19. I remember best:
a) What I see
b) What I hear
20. It is more important to me that an instructor:
a) lays out the material in clear logical steps
b) gives me an overall picture and relates the material to other subjects

Example questions from the Felder Test.

If you would like to determine your own learning style, the Felder Test is attached as Appendix A. Try it. You will enjoy it.

Back in the UK we presented the four modules to students on the HCI course, and we added the learning style test before the presentation of the first module therefore, all students were asked to answer the 44 questions in the Felder questionnaire to determine their learning style.

All students really enjoyed taking this test and they were interested in their individual result. The results across the students as shown. As expected, the visual style was much more common than the verbal style, so this axis was ignored in these experiments.

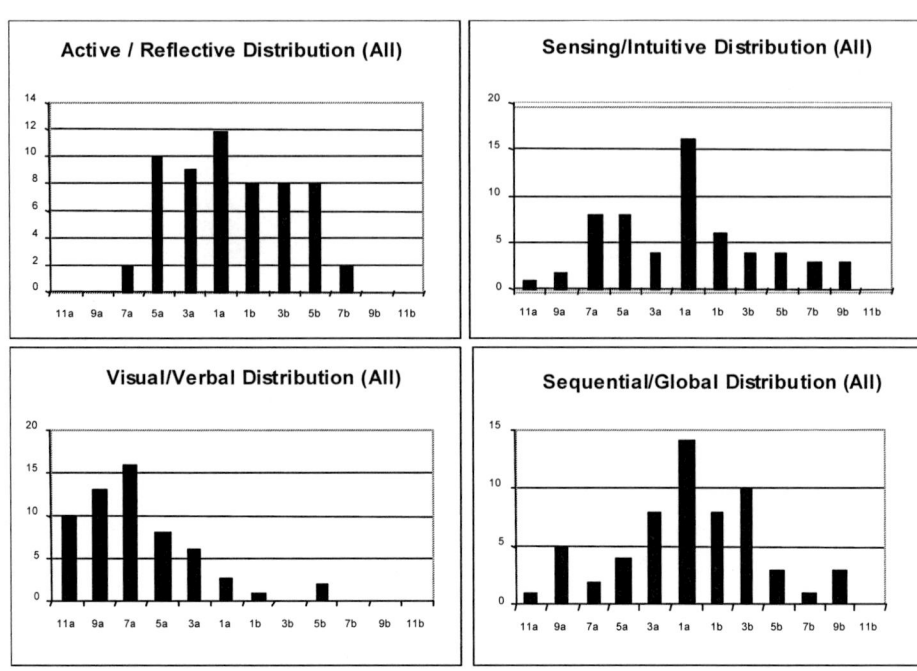

The Student Learning Styles Distribution

The nature of the Statistics domain did not lend itself to animation. However, ideas and diagrams were built up progressively on the screen. As new diagrammatic elements were introduced, the text would simultaneously appear on the screen, or the spoken commentary would occur. Occasionally, blinking was used to emphasise elements being discussed. Colour was also used to connect important sections of text or diagrams.

The final tests were conducted immediately after the presentations. The groups of students were moved between the presentation formats on succeeding days to avoid biasing the results. It was important, of course, to check if students already knew about Statistics. In fact, few students had previous knowledge of the subject area, and even some of the professed knowledge was incorrect. Students with more than 30% previous knowledge were eliminated from the test, but only one was actually eliminated. No students were eliminated on the third and fourth days. Altogether there were 61, 66 and 66 students in each of the three presentation types

The vertical axis shows the learning score obtained (between 1 and 13). The most striking result was the superiority of the Sound + Diagrams presentation (in black) format over the other two scores. The scores achieved in all the four modules as shown.

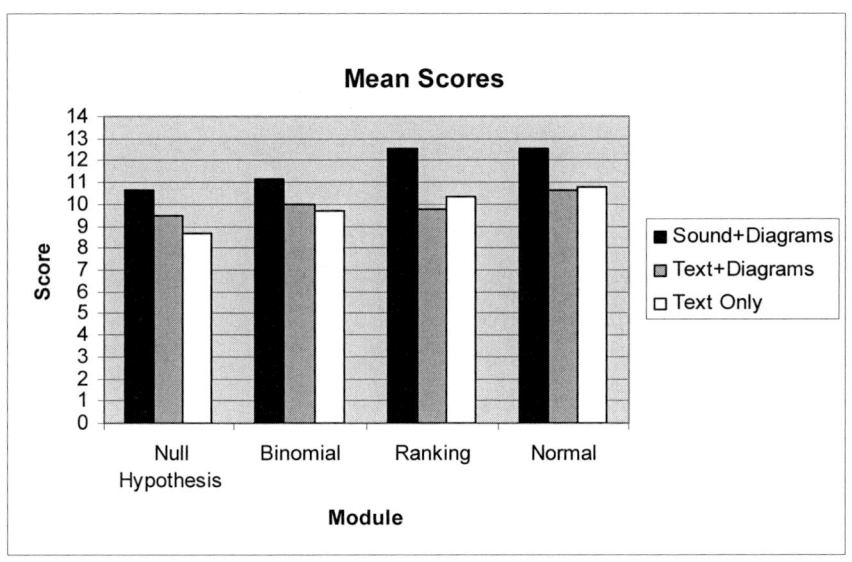

The Results of the Test

This is a very significant result. For readers who understand Statistics, a one-way ANOVA analysis on performance between the three presentation styles revealed a significant difference (F=4.612, 2, 190, p<0.011).

Furthermore, the effect persists across the different module contents even though the nature of the content varied considerably. For example, the first module (Null Hypothesis) is very descriptive, whereas the Binomial and Ranking modules are much more mathematical in nature. Although scores generally increase over the four days there was no appreciable learning effect.

The results agree with other workers. A similar improved performance for the Sound + Diagrams presentation is observed as with Mayer's Sound + Pictures presentation. The similarity in performance between Text + Diagrams and Text Only surprised us. Theory (and Mayer's results) predicts that the former will be more effective.

An analysis was then carried out to determine if the participants' **learning style** had an effect on learning. There were no clear effects for Global versus Sequential learners or for Active versus Reflective learners. However, there were interesting differences for Sensing versus Intuitive learners.

The Intuitive learners performed better overall, and it was not clear why. It is not due to the type of presentation nor the information in each module. However, the content of the four modules is more theoretical than practical and this might explain the result.

Unknown to me, Nigel had added a question to the student profile asking if students were Dyslexic and it turned out that there were six registered Dyslexic students in our sample. We were therefore able to examine the difference in learning scores between Dyslexic students and non-Dyslexic students. Although the sample was too small to achieve any significance, the experimental results suggested that computer-based media combinations might affect learners who have Dyslexia differently to non-Dyslexic learners. This was unexpected, since the learning materials used involve both verbal and nonverbal content.

Dyslexia is defined in the following way

Dyslexia is a severe difficulty with the written form of the language independent of intellectual, cultural and emotional causation. It is characterised by the individual's reading, writing and spelling attainments being well below the level expected based on intelligence and chronological age. The difficulty is a cognitive one affecting those language skills associated with the written form, particularly visual, verbal coding, short-term memory, order perception and sequencing.

We therefore designed a new experiment using 30 Dyslexic students from Loughborough University. The participants were taken from various courses taught at the University and all volunteered for the study. Participants were mainly from science departments (10), and engineering departments (12), but there were 8 students from arts departments. To ensure that the participants really were Dyslexic their degree of Dyslexia was assessed by two tests—The Lucid Adult Dyslexia Screening Test (LADS) and a Visual Perceptual Problems Inventory (VPPI).

Because of the different backgrounds of the students in this experiment, each participant was given a pre-test to check their previous knowledge. The three groups were then presented with the material from the first module (The Null Hypothesis module) in the three different media combinations—sound and diagrams text and diagrams, and text alone. Then, after seeing the presentation, each participant was given a post-test. The results are as follows.

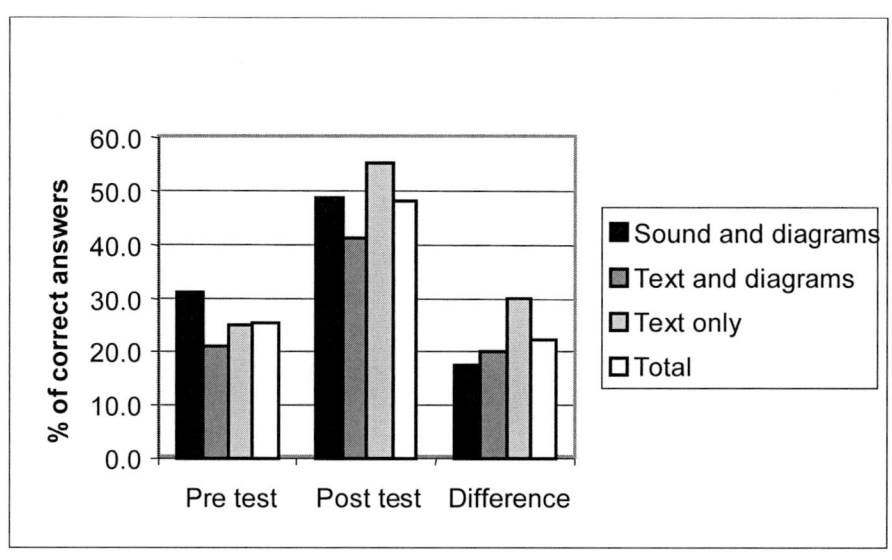

Performance on the Test

If this is compared with the earlier results, it is interesting that the Dyslexic students responded quite differently to non-Dyslexic students. Text-Only had better learning scores and there was a significant difference between Text-Only and both Sound + Diagrams. A full report of the experiment is given in a paper by Beacham and Alty, (2006).

The Dyslexic students came from a wider population than the students in the previous experiment. The students in the former experiment were all computer literate students who had an interest in Computer Science. In the Dyslexic student group two thirds of the subjects were Scientists or Engineers, one third of the students came from Art and Design, Social Sciences and Business Studies.

The results from the study suggest that the combinations of media affected the understanding achieved by Dyslexic and non-Dyslexic students in different ways. For example, the media combination Sound + Diagrams gave significant performance difference for non-Dyslexic students, but not for Dyslexic students. In contrast, Text-only presentations had the opposite result—Dyslexic students performing better with this media than with the other combinations, and this wasn't expected. Whilst we would have expected the Dyslexic subjects to have problems with text alone, it was not expected they would have problems with text and diagrams or with sound and diagrams. Interestingly, some Dyslexic subjects obtained higher scores when having information presented as text alone than with text and diagrams.

When we presented our findings at a major Conference in Hawaii, it aroused a huge amount of interest, and we received the best paper award of the conference! The paper has been (and still is) highly quoted, and I think it was some of the most satisfying work that Nigel and I were involved in.

The study suggest that information presented using text and diagrams for non-Dyslexic learners may not be the best way of presenting information to Dyslexic learners. This needs further study.

In 2000, the University asked me to become Dean of Science. I did continue my interest in research but at a reduced level. In 2003 I retired but continued to work part-time advising student on their final year projects and lecturing on HCI courses. I finally fully retired in 2011. It had been a great career but not really planned!

Chapter 26
Looking Back

So many events in life happen by chance and the trick seems to be to keep in mind what appeals to you and be ready to seize any chance quickly if it happens! I have never had a very fixed view of what I should do in the future, and most things seem to have happened by accident. Getting into Physics and Mathematics was the result of a chance discussion with Peter Smith as we walked homewards across the Golf course. I carried on into the sixth form because I couldn't think of anything else to do! My intense love of Classical Music happened by chance when I borrowed the 78-inch record of the first movement of Grieg's Piano Concerto, again from Peter. Frank Duckworth gave me an interest in Theorems and although my first encounter with Differential Calculus was a disaster, my second encounter blew me away in wonder.

I only received one first class honours (and the Oliver Lodge Prize for the best student) whilst at University—but it was the one that mattered, and this was partly caused by a disastrous love affair of one year ended six weeks before finals and this made me work my socks off!

My move into Nuclear Physics was not initially my first choice. I had intended to do a PhD in Metallurgy, but when I told the Head of the Physics dept of my desire, he exploded and said, "the top student in physics is not going to do Metallurgy!" and I didn't dare to contradict him. However, it was the right choice. A PhD in Nuclear Physics gave a student experience in Computing, in Mechanical Workshops, in Electronics and high-level Mathematics, and in Group Working.

The move to IBM turned out to be an inspired choice, but it was really motivated by a disagreement with the Head of Metallurgy. I loved the technical side of IBM and became a Senior Systems Engineer, but IBM pushed me into

Sales. I was not sure that I would be good at Sales, but it turned out to be an interesting job and taught me a lot.

Then, just as I had become top student in the Sales School at IBM, and people told me that my career was made in IBM, out of the blue I received the offer of Head of Computing at Liverpool University. IBM tried to persuade me not to leave, pointing out that Liverpool had fallen behind in Computing, but I thought it was too good an opportunity to miss so I left.

Although I knew University life well, I had never really managed anything. It was the first time I had a full-time secretary and on the first day she came in and asked me what she should do next. I replied, "I don't know but I am sure something will turn up!"

On arriving at Liverpool, I found that the Computer Centre had been badly run, and the Computer Board had downgraded the computer replacement, insisting that half the load should be obtained from Manchester. Initially I was not sure how to handle this problem, but my sales training in IBM helped. The Board has said that they expected 50% of the load to be offloaded to Manchester and the Computer Centre had been discouraging users from submitting jobs. I suddenly realised that the opposite was the correct response. Encourage the users to submit jobs to Manchester and build up such a huge load that 50% of it would require a much bigger computer at Liverpool—and it worked!

However, the side effect of this policy was much more important. I realised the importance of making computers easy to use, and a life-long interest in Human Computer Interaction was born.

When Microprocessors came on the scene, it was my worry about them making mainframes obsolete (and affecting my job) that caused me to take a keen interest in them (rather than an interest in the technology!). However, they did fascinate me, and we set up the Liverpool Workshops. I also set up the first major Microcomputer Laboratory and the Microcomputer Working Party Report, which I chaired, and this had a major effect on Universities and Industry.

Liverpool was a great success, but I needed a change and moved to Strathclyde. Again, an unexpected event happened—the moving of the Turing Institute to Glasgow. I had been dabbling in AI with Mike Coombs and when Sir Graham Hills asked me to take over the Institute, I wasn't sure if it was the right move. However, it turned out to be a great move and I met many interesting people. In setting up the Scottish HCI Centre with Heriot-Watt University I realised how difficult University politics can be!

Although I enjoyed working in AI, HCI was my main interest, so moving down to Loughborough to join a strong HCI Research Group (run by Prof Edmunds) was a good idea. A chance meeting, with a senior person from CRI in Copenhagen, resulted in me becoming involved in Process Control and working with European colleagues in a major European Research Project. This was really interesting and over the next 14 years I became heavily involved in four large European Research Projects.

Another chance event happened when Iain Duncumb came to work in the Research Group at Loughborough. He saw the opportunity of working in DAB (I hadn't even heard of it!) and as a result we worked with Roberts Radio to produce the first portable DAB receiver (one of which is now in the British Science Museum) and the subsequent cheaper version produced by Roberts made DAB a reality.

Finally, Nigel Beacham coming to work with me in the mid 90's had a major effect. As I explained earlier, he was Dyslexic and when we were examining how presenting material in different combination affected Student Learning he, on his own initiative, looked into how Dyslexic student reacted to the different media combinations. This really impressed me, and we carried out a major study of Dyslexia, which is now widely quoted, and I hope had some effect on improving the understanding of Dyslexia.

I said at the beginning, much of life seems to happen by chance, and that is certainly true in my case. I sometimes envy those who seem to be able to plan their future before them, but too much planning can make you blind to opportunities which arise. I suppose that the key issue is whether one sees an opportunity in a lucky (or unlucky) event. Einstein once said that "in the middle of a problem lies an opportunity" and the Chinese word for crisis contains the two words meaning "danger" and "opportunity". Should I have left IBM for Academia? Should I have become the Director of the Turing Institute when I didn't know much AI. Should I have taken on the senior job of Systems Engineer at Daresbury before I had completed my Training? We will never know—and it probably doesn't matter!

I suppose, in reality, flexible planning is the real answer!

Chapter 27
The Digital Age:
The Good and the Bad

We have come a long way since Alan Turing first wrote his famous paper and heralded in the Digital Age. He would have been over 100 years old today if he had lived. The development of the Universal Turing Machine changed forever our ways of processing and communicating information and it is interesting to note that he is now better celebrated than ever before, and a new film about his life was released to acclaim. A major new Institute has been proposed by the Government which is likely to be called The Turing Institute. The previous Turing Institute in Glasgow, which I ran for a number of years, later closed its doors and ceased to trade in 1994, about 4 years after I left.

In recent times our attitude to gay people has also improved. Alan Turing was given a royal pardon by The Queen for his 1954 conviction.

But how has the Digital Age changed things? Let us examine the effects of the Digital Age. Both good and bad.

27.1. A Huge Technological Change has Resulted in Major Changes in the Way We Process Information.

The changes have been dramatic particularly in the way information is processed. On the first computer I used in 1963 (DEUCE) the computer had 16 fast stores and a high-speed drum. It could carry out a calculation in 64 microseconds. When I was Director of the Computer Centre at Liverpool in 1972, the KDF9 computer had 128 Kbytes of main store and could do a calculation in 6 microseconds. The 1906S installed at Liverpool in 1995 had over a megabyte of main store, 6 x 60Mbyte disks and supported multiprogramming. It had a cycle time of 250 nanoseconds. The availability of main store continued to rise (in what is often called Moore's Law) and became much cheaper. Today 100+ Gigabytes is not uncommon even on small computers or a household television. The speed of calculation has exploded in a similar fashion.

Availability of fast networks is now commonplace—a big change from the 1.2 Kbytes/sec which was available in the early 1990s.

The Digital Age has changed everything about our lives. Firstly, the way we have processed and communicated information has changed beyond all recognition. Information is communicated much faster, more accurately, in much greater volume, The Internet has connected us all, and has changed the way we talk to each other, the way we often shop, the way we book holidays and bank our money, and even the way we learn. The number and composition of the shops on the high street are undergoing many changes as a result of the threat of on-line shopping.

The technology has resulted in many interesting devices which are really useful. For example, an App can be installed which shows the user where they are in any location. In a city, the way can be determined to any nearby location, and in the countryside, it can tell the user where they are and what towns are near. The Satnav has made driving easier. A recently released App tells the user where hot spots are in the house or garden. It can find a pet, check if the sausages are properly done, or where best to plant the strawberry plants! Other heat seeking Apps are used in rescue operations when people are trapped in fallen rubble. There are many Apps which tell the user how healthy they are.

We can access lots of Apps to do jobs for us and many of them are free or cheap. We can now translate one language into another with very little knowledge of the target language. We can spell check our writing and even write music on the computer. However, these advantages sometimes come with a price. People are now probably worse spellers that ever before, and on-line translation whist often very effective, can occasionally give odd results. The idea of "Driverless cars" is now being trialled, but people have been killed e.g. when a driverless car crossed a junction and drove straight into a large lorry, and when a driverless car failed to recognise a woman on a bicycle.

Although the user now has a vast knowledge base at his or her disposal and can find out almost anything at the touch of a button, they often are not sure how correct the information is. Certainly, knowledge bases like Wikipedia are good quality knowledge sources. People can access the knowledge and suggest changes to it when they feel is incorrect, but this can also be challenged by other readers. Overall, this works quite well.

However, other knowledge sources are of varying quality. Some are often out of date. People are good at putting up new information on the net but not so good at taking down out-of-date information. Another problem is "Fake News". Frequently articles appear on the net which have no foundation in the truth whatsoever. Later they are repeated in newspapers and magazines. There were claims that the election of Donald Trump was influenced by fake news, some claimed, generated by Russian sources. Facebook recently reported that about 13% of the accounts on Facebook were either duplicate or fake, they estimated that fake accounts were about a fifth of the 13% (some of which had been created for "spamming" or other malicious purposes).

Another worrying trend is the sharp increase in "fake reviews" on the net for products. Firms now exist which pay people to fake 5* reviews on products on the net (for example on Amazon). If a review is "too good to be true" it probably is!

Even when the knowledge transmitted is correct it can cause problems. Many authors are now very concerned about copyright. Copyright is the main mechanism by which they are rewarded for their hard work. Sites have now been set up whereby anyone can download free copies of their books and the same is happening to music. These sites are illegal, but no one seems to know how to deal with the problem. The owners of the sites claim that "knowledge should be freely available with no restrictions" but curiously they do not allow others to have private knowledge about them!

Forecasting the future has always been fraught with difficulty. The problem is that one can never predict what technological changes will take place which makes your thinking obsolete. Think of London in the 1870s. Planners became very worried about the amount of horse manure being deposited on the streets. They predicted it could become feet deep! Yet within 20 years the invention of the motor car made such thinking obsolete. The same is true today. Who could have predicted the full effect of smart mobile phones in the 1970s? Our whole shopping experience has changed, and information is spread so rapidly it is now sometimes affecting our capability for making good rational decisions. It is not the changes that are the main problem, it is because they change too fast.

Human beings are capable of absorbing new environments. The introduction of automation and new forms of transport (car, aeroplane, telephone) have had huge effects on the way we carry out our lives, but they were introduced at a reasonable pace. We had time to check and adjust to the good and questionable effects. The problem with digital technology is that it is being introduced at a breath-taking pace and we do not have the time to evaluate the effects and adjust our habit and laws to accommodate it. There are major issues which are not being properly addressed – privacy versus openness, open access versus security, effects on vulnerable people, misleading and fake information, and the development of the Dark Web.

The way people assess the importance of technology is interesting. There is a law called "Amara's Law". This suggests that whenever a new technology arrives, human beings overestimate the possible effects in the short term but underestimate the effects in the long term. It seems that we tend to get it right about 15 years down the line.

27.2. Artificial Intelligence

Artificial Intelligence has had a chequered history. In the past, AI workers have often tended to "over-egg the pudding". One consequence of this was the Lighthill report which criticised the effectiveness of AI research in the late 1950s. Research workers had promised too much and there was a backlash. There is a tendency for AI to still do this today. Now every implementation is claimed to be an AI success, and there are many comments about AI robots taking over! However, it depends on what is meant by "Intelligence".

When AI claims success, like when "Big Blue" beat Kasparov at Chess, it is not strictly correct. Certainly, much chess knowledge was contained in the IBM programme, but there were also factors which indicated that comparing the intelligence of the computer with the intelligence of Kasparov was not strictly correct. The computer was using techniques which are not present in the human brain. For example, A computer can look 100's of moves ahead in its planning whereas a human being can only look a few moves ahead. So, part of its success was based upon a technique which are not present in the human brain. A better term for "Artificial Intelligence" is probably "Machine Intelligence".

Computers have always been able to outperform human beings in recalling information or doing complex calculations, but it does not mean that they have superior intelligence. I also question why AI workers are so obsessed with "human-like" robots. Why have a robot with 2 arms and legs and two eyes, when a robot could have 8 legs and 6 eyes with two in the back of the "head"? I am not suggesting that AI is not useful. Valuable progress has been made, particularly in areas of capturing narrow expertise. Maybe I am at the stage (as in Arma's Law) where I am disappointed by the gains over the past ten years but could be shown to be wrong in 20 years' time.

Of course, digital technology exhibits the same characteristics as all other human developments. As Adam and Eve found out when they tasted the tree of knowledge, acquiring knowledge can have both good and bad effects. It seems that for every benefit there is often a drawback. This is equally true of the new technology.

27.3. Response to Customers

The Internet may offer us quicker and easier access to services but that only works well when the customer knows what they are wanting. In the 1970's, if a customer had a query, they could ring the supplier and get an immediate response. Today it is difficult to find the right telephone number to ring, and even more difficult to find a contact telephone number on the company web site. This seems deliberate. If there at all, such telephone numbers are often hidden away in an obscure place.

But even if a customer finds the number, they can be waiting many minutes for a phone call to be answered. "All our customer service staff are busy, please try later" is the normal answer from most suppliers these days and it happens whatever time you try to connect. It is not just at busy times. It seems that suppliers are deliberately trying to discourage customers from telephoning them, and their "Frequently Answered Questions" on the website never answers any important questions. I have just spent 35 minutes on the telephone solving a problem which in the old days would have taken 5 minutes!

On the other hand, some suppliers are also obtrusive. They will bombard a user with information that the user may not want. There is often the "unsubscribe" facility, but this is sometimes hidden away at the end of the website. The way of communication with the user has also deteriorated. In earlier chapters, the reader has seen the real improvements that were made in Human Computer Interaction between 1970 and 1990. Unfortunately, this has not always fed through into website interfaces. Today, website designers are often repeating the same mistakes that the earlier computer interface designers made!

27.4. The Internet of Things

This is a new term to describe the way almost every device is now being attached to the Internet—home devices, wearable technology and cameras.

27.5. The Problems of Personal Contact

The revolution in telephone technology—the development of the Smart Phone—has changed the way we communicate, particularly for the young. Parents now provide their children with a smart phone, and this has the advantage that they can keep in touch with their loved ones, and this is much safer than before. However, there is evidence accumulating that communication has changed between young people not always in desirable ways. For example, "cyberbullying" has become much more common between young people. Services such as Twitter have created an audience of uncritical, irresponsible, instant commentators on matters of public interest. This new audience often does not allow an opinion which differs from the "acceptable".

People are not allowed to make a simple mistake when expressing an opinion and are hounded out of jobs because of one slip-up, over saying something which they did not intend to be taken in a particular way, even though their recent history supports the alternative view. No account is taken of the changing moral view of society. Fifty years ago, some remarks would have been regarded as normal comments but today they are unacceptable. So past remarks should be judged by the standards of the day they were uttered, not those of today. We seem to have forgotten how to forgive.

Young people are becoming addicted to their phones. On recent survey claimed that many young people would rather forgo sex than have their smart phone confiscated from them! In another survey of students, they were asked if they would prefer a broken phone or a broken bone, 46% replied they would prefer a broken bone! Currently, 70% of office emails are read within six seconds and 60% of replies are sent within 1 hour. Recently Apple confirmed that users unlock their phones 80 times a day. Another study showed that a typical user touched their phone on average 2617 times a day. Nomophobia—the fear of being out of contact—has become a psychological problem.

Good Manners also seem to have disappeared. Smart phones will come out and be used in public meetings, when people are talking together, or even when they are having lunch or dinner. In many cases the smart phone has taken over as the main form of social interaction. One problem with this sort of interaction is that it does not communicate fully what people are feeling. In a face-to-face conversation the recipients get a lot of feedback on how their comments are being received. A slight raising of the eyebrows, or a movement of the mouth can communicate displeasure at what has been said. This does not happen in a smart phone conversation. The result is that people say riskier and possibly objectionable comments, but do not realise that they may have caused offence.

Addiction also causes accidents. People now walk along typing into their phones on the high street, or even when crossing the road. A recent newspaper article in the Times reported that the National Help Line has reported that half of 16 to 24-year-olds have walked into an obstacle whilst checking their mobile phones, and the number is still 13% across the whole population! In the United States, a rise in pedestrian deaths and thousands of extra injuries have been blamed on texting whilst walking. A sculptor had to move a statue outside Salisbury Cathedral, because so many clumsy walkers were colliding with it whilst texting! In Sweden signs have been posted warning texters and phone checkers, and in California signs have been added to road crossings "Heads Up! Cross the Street, then update Facebook"!

Even worse is the use of mobiles whilst driving. There already have been a number of fatalities caused by drivers being distracted using mobiles whilst driving. Of 88 deaths in 2012 caused by distraction of attention, 17 were due to mobile use. As I write a woman has been jailed for 5 years for causing a serious accident on the M1 which killed one person and seriously injured another. She was using her mobile phone when she lost control, and then tried to delete the mobile log. Yet many people are still using mobile phones whilst driving.

These days there is very little that anyone can say on the web without someone taking offence. There seems to be a large band of people who spend their time taking offence at anything being said. Such people often do this without checking if any offence has been caused! One Christmas recently, a headmistress banned the Christmas Nativity play because it would cause offence to Muslims. As a result, she received complaints from Muslim parents that they wanted their children to take part and having it would cause no offence. She hadn't bothered to check.

Many actions of politicians these days are attacked on Twitter by an army of ill-informed complainants who react immediately without considering the full arguments. Of course, politicians do get things wrong, but the rebuttal should be formed from a proper consideration of the facts not a knee-jerk reaction.

I am sure that human beings can solve many of the problems outlined above. But they need time to do this.

27.6. Loss of Attention

It is well-known that the attention span of children is reducing. Some calculate that it has dropped quite considerably. You only must look at TV advertisements to see how the attention span is dropping. Frequently, cartoon characters are now used in advertisements for adults, and promotional material on television now requires 30 seconds of quickfire flashing pictures to get over a basic message. "What the Papers Say" on the BBC is now advertised with a 30-second video of rapidly changing pictures whereas the simple message would probably take about 4 seconds.

Some recent experiments comparing reading printouts with observing information have shown that people tend to concentrate on the detail and overlook the broader picture when using a screen in comparison with those who see the same information on a printout. The researchers compare the effect with "tunnel vision". In an experiment where the details of four cars were examined on a printout and compared with the same information on a PDF displayed on a screen, 66% of those using the printout correctly identified the best car, compared with 34% for those using the screen. Another experiment concluded that people using the IPAD did not think about the long-term implications of their decisions.

27.7. The Handling and Protection of Our Finances

Our money may be easier to access and process at the bank, but so it is for unscrupulous people who wish to take it away from us. One estimate in 2016 was that over £26 billion had been wrongly extracted from people, often from the old and vulnerable. We have lost the close connection we used to have with our bank. We now receive emails asking us for action whose source is difficult to check. We receive phone calls which ask us to return the call, but the actual return call is to a different number than implied. The banks, of course, have put in place security mechanisms such as passwords protection and stress to customers that they will never ask for a password, but the criminal classes often use the technology to get around these measures. Older people particularly are not well versed in the details of the technology and can easily be misled by unscrupulous people.

The widow of a friend of mine recently was approach on the telephone by a plausibly sounding lady who, over the following week, gradually got her confidence by claiming she was from the bank investigating suspicious activity on her account. The widow thought that by ringing back to the calling number (which looked like the bank number) she was talking with bank staff. The caller pointed out that the criminal activity might even be in the bank itself so that the widow should not communicate with other bank personnel! The criminal finally persuaded the widow to part with her account number and password and the widow lost about £14,000. The bank refused to compensate her, and so she lost the money.

The banks and other financial companies claim that if a customer follow their procedures that your money is safe, but recently there have been a series of "Cyber Attacks" on accounts where sensitive private information has been obtained despite the protection mechanisms in place. The network providers have not properly addressed the access and protection issues.

Of course, it is not surprising that the internet has created much fraudulent and dishonest activity. Criminals have always taken advantage of new developments. There are so many of these that it is difficult to cover them all. Fraudsters will set up realistically looking web sites to tempt users to give up vital information. The web sites will look like the genuine article. I had one recently telling me that I had a £230 tax rebate owing to me. It asked me to contact the site (with my bank details) to obtain the refund. When I checked it closely it was clearly a fraudulent site. The email return address was wrong and many of the additional buttons on the site did not work when pressed (these two actions are usually quite effective at discovering if a site is genuine). I contacted the Inland Revenue and of course, the site was fraudulent.

Other fraudsters send users demands for payments which the user does not owe. Also, a disturbing business of swapping email addresses illegally has become commonplace. This is often called "Phishing". This then enables the fraudsters to contact you (often having also extracted your mailing list) purporting to be, perhaps, one of your friends.

As I write this chapter, I received an email from a good friend of mine who I have mentioned a number of times earlier in the book—Frank Duckworth. The email looked genuine (i.e. it was the correct email) and it was asking me to assist him with a problem. I was immediately suspicious because it did not use my first name and it was very brief. It claimed to be concerned with his niece who had cancer! I replied asking for two pieces of information that only Frank would know—and I got no reply! The real concern is that my reply was to the valid email, but it went to the Fraudster!

I also recently had an email from a "good friend" of mine claiming that he was stranded in Greece and asked me to send £150 to assist him to come home! Fortunately, I managed to contact him and although it was true that his mobile phone had been stolen, he did not require money! Often with such emails the fraudulent nature can be easily exposed if you look carefully at the content. They often use the wrong Christian name or make stupid English spelling errors.

Of course, criminals will always take advantage of new developments. There is nothing new in this. It has happened for thousands of years, but the pace of change means that new fraudulent activity changes its nature very quickly and people (particularly older people) are slow to recognise it (and it takes a long time to warn them).

27.8. Viruses and Protection

The user receives a phone call saying that their computer has been hacked. The caller offers to put the situation right. The callers are chancers who have lists of thousands of UK phone numbers. Some profess to be Microsoft and offer to clean your computer for a fee. However, even Microsoft cannot know if your computer is having problems and then associate these with your landline and phone you. In May 2017, a virus called Wannacry caused serious outages in well-known organisations such as the NHS, and it even got past well-known protection software for individual users. A well-known package which claimed, "Total Protection" did not prevent Wannacry from entering the computer and freezing the files.

Should users therefore install Virus Protection Software? Yes, but realise that the small print often points out that they are not responsible for any damage caused by viruses if their protection fails. Realistically a good anti-virus package will protect a user from most threats, but they are not totally invincible. In a sense, they are always catching up.

27.9. Hackers

There are computer people today who spend their whole time trying to access private information. They are called "Hackers". Some do this for intellectual satisfaction (they are not malicious; they are just seeing what is possible). Others advise large firms on how to better protect their information. But some Hackers are very malicious and try to hack into bank accounts or other large databases (such as the NHS).

There are annual conferences each year held at Las Vegas (Called DEFCON) at which Hackers show off their prowess. Up to 10,000 people a year attend this conference. They show how they can take control of laptops, cameras, printers, routers and most of the devices used in "smart" houses.

Their motives may be intellectual, political, or simply malicious.

27.10. Terrorists

Terrorists have found the web extremely useful in getting their message across. One worry is that terrorists may soon be able to create biological weapons, the EU counter terrorist co-ordinator said in 2017 that this may follow on from the ISIS activity. For example, "how to make a bomb in your mum's kitchen" could evolve into "how to process a virus in your mum's kitchen".

27.11. Improper Use of Video Material

Many people put videos of amusing incidents or important events on YouTube. Most are harmless, and some are intended to be harmless but can have unfortunate consequences. These videos can be viewed millions of times and they are therefore a prime site for advertising. YouTube charges for these adverts and many large and important businesses use these highly viewed videos to advertise. Currently, the creator gets 55% of revenue and YouTube gets the rest. A creator can get £7 per 1,000 views, so a video viewed 4 million times can yield £28,000 for the creator. People can also make comments on the videos.

Big companies like Google BT, Adidas, Deutche Bank, eBay, Amazon, Mars, Stella McCartney, Argos and Dolmio are all advertising on these Video Sites, but they often do not know which video sites are showing their advertisements. Such sites can even include terrorist sites. The sites are chosen because they have a huge following, but the content may be very questionable. The providers are not doing enough to stop this happening. Again, there has been too little time for adjustment.

Recently Governments are becoming involved, and it is expected that much more stringent regulatory policies will be developed.

27.12. Conclusions

Was my career worthwhile? Compared to Alan Turing's enormous contribution, my contribution is trivial! At Strathclyde I knew the Principal, Sir Graham Hills really well, and I thought he was a great Vice Chancellor. When it was announced that he was leaving, I started a campaign to keep him on for a further period and to my surprise met considerable opposition. I told Sir Graham and he replied "Jim, don't ever expect to be rewarded or thought well of because of what you have done. Whatever you have done it will not usually be recognised. The key issue is how you view your efforts. If you are satisfied that is all that matters!" Ironically, I now realise that most of the good things I did happened through chance. Whether we like it or not Luck plays a large part in our lives.

I have probably been perhaps too critical of the current state of the Digital Age. The web has been an amazing invention and it has transformed most peoples' lives. Wikipedia will probably replace Encyclopaedia Britannica, and language translation has dramatically changed. Shortly we may have the equivalent of the Babbel Fish for instant translation as in the book—The Hitchhiker's Guide to the Galaxy.

Who would have thought that so much change could have happened so quickly? Human beings are usually capable of dealing with major changes to their life and environment, but they need time both to understand the effects of change and to make the adjustments. New devices and applications are now flooding out with little concern about the good and bad effects of such applications.

I think human beings will be able to eventually sort out the undesirable effects. But we do need time to think about it. All human inventions have benefits and drawbacks, and time is needed to adjust our approach so that the benefits are realised, and the drawbacks minimised. For example, saying that all knowledge should be openly available and free of charge sounds a good idea but at the same time it compromises both personal privacy and national security.

Perhaps in the fullness of time we will discover how to use the internet well and limit the undesirable effects. Would Alan Turing be worried by recent trends? Is technology sending us hurtling into a new dawn or a nightmare? Who knows!

Bibliography

1. *Alan Turing: The Enigma* by Andrew Hodges, published by Vintage Press. Any reader who wishes to know more about Alan Turing should read this.
2. *Early British Computers* by Simon Lavington, published by Digital Press.
3. *Computing Perspectives* by Maurice Wilkes and Morgan Kaufman.

For computers and music:

4. *When Bugs Sing,* Vickers, P. & J.L. Alty, Interacting with Computers, 2002, Vol 14(6), 793-819.
5. *Communicating Graphics via the Auditory Channel, An Empirical Approach,* Int. J. Human Computer Studies, Vol 62, pp 1-10.
6. *Miller*

For the gradient and promise projects:

7. *Knowledge Based Dialogue for Dynamic Systems,* Alty, J.L., and Johannsen, G., (1989), Automatica, Vol.25, No. 6, pp 829-840.
8. *Knowledge Engineering for Industrial Expert Systems*, G. Johannsen, and Alty, J.L., (1991), Automatica, Vol. 27, No. 1, pp 97-114.
9. *The GRADIENT Dialogue System; Providing better interfaces for Process Control,* Alty, J.L., (1989), in Computer Technologies, (Salieneks, P. Ed,), Ellis Horwood, Chichester, pp 75-101.
10. *The Design of the PROMISE Multimedia System and its Use in a Chemical Plant,* In Multimedia Systems and Applications, (Earnshaw, E., and Vince, J.A., Eds.) Academic Press, pp 53-75.

DAB:

11. *The Role of Media Efficiency during Learning Using a DAB Radio,* Proc. of Ed-Media (2004), (Kommers, P., and Richards, G., Eds.), Lausanne, Switzerland, pp 4309-4316.

Computer aided learning:

12. *An Investigation into the Effects that Digital Media have on the Learning Outcomes of Individuals who have Dyslexia (2006),* N.A. Beacham, & J.L. Alty, Computers and Education, Vol. 47, pp 74-93 (2006).
13. *Dual Coding Theory and Education: Some Media Experiments to Examine the Effects of Different Media on Learning (2002),* Alty, J.L., In Proc. of ED-Media (2002), World Conference on Educational Multimedia, Hypermedia and Telecommunications, Denver, Colorado, Keynote, pp 42-47.

Appendix 1
The Felder Test

For each of the 44 questions below, select either "a" or "b" to indicate your answer. Please choose only one answer for each question. If both "a" and "b" seem to apply to you, choose the one that applies more frequently. Remember, there are no "wrong" answers.

1. I understand something better after I
    a) try it out.
    b) think it through.
2. I would rather be considered
    a) realistic.
    b) innovative.
3. When I think about what I did yesterday, I am most likely to get
    a) a picture.
    b) words.
4. I tend to
    a) understand details of a subject but may be fuzzy about its overall structure.
    b) understand the overall structure but may be fuzzy about details.
5. When I am learning something new, it helps me to
    a) talk about it.
    b) think about it.
6. If I were a teacher, I would rather teach a course
    a) that deals with facts and real-life situations.
    b) that deals with ideas and theories.
7. I prefer to get new information in
    a) pictures, diagrams, graphs or maps.

b) written directions or verbal information.
8. Once I understand
   a) all the parts, I understand the whole thing.
   b) the whole thing, I see how the parts fit.
9. In a study group working on difficult material, I am more likely to
   a) jump in and contribute ideas.
   b) sit back and listen.
10. I find it easier
    a) to learn facts.
    b) to learn concepts.
11. In a book with lots of pictures and charts, I am likely to
    a) look over the pictures and charts carefully.
    b) focus on the written text.
12. When I solve math problems
    a) I usually work my way to the solutions one step at a time.
    b) I often just see the solutions but then have to struggle to figure out the steps to get to them.
13. In classes I have taken
    a) I have usually gotten to know many of the students.
    b) I have rarely gotten to know many of the students.
14. In reading nonfiction, I prefer
    a) something that teaches me new facts or tells me how to do something.
    b) something that gives me new ideas to think about.
15. I like teachers
    a) who put a lot of diagrams on the board.
    b) who spend a lot of time explaining.
16. When I'm analysing a story or a novel
    a) I think of the incidents and try to put them together to figure out the themes.
    b) I just know what the themes are when I finish reading and then I have to go back and find the incidents that demonstrate them.
17. When I start a homework problem, I am more likely to
    a) start working on the solution immediately.
    b) try to fully understand the problem first.
18. I prefer the idea of

a) certainty.

b) theory.

19. I remember best

    a) what I see.

    b) what I hear.

20. It is more important to me that an instructor

    a) lay out the material in clear sequential steps.

    b) give me an overall picture and relate the material to other subjects.

21. I prefer to study

    a) in a study group.

    b) alone.

22. I am more likely to be considered

    a) careful about the details of my work.

    b) creative about how to do my work.

23. When I get directions to a new place, I prefer

    a) a map.

    b) written instructions.

24. I learn

    a) at a fairly regular pace. If I study hard, I'll "get it".

    b) in fits and starts. I'll be totally confused and then suddenly it all "clicks".

25. I would rather first

    a) try things out.

    b) think about how I'm going to do it.

26. When I am reading for enjoyment, I like writers to

    a) clearly say what they mean.

    b) say things in creative, interesting ways.

27. When I see a diagram or sketch in class, I am most likely to remember

    a) the picture.

    b) what the instructor said about it.

28. When considering a body of information, I am more likely to

    a) focus on details and miss the big picture.

    b) try to understand the big picture before getting into the details.

29. I more easily remember

    a) something I have done.

    b) something I have thought a lot about.

30. When I have to perform a task, I prefer to
    a) master one way of doing it.
    b) come up with new ways of doing it
31. When someone is showing me data, I prefer
    a) charts or graphs.
    b) text summarising the results.
32. When writing a paper, I am more likely to
    a) work on (think about or write) the beginning of the paper and progress forward.
    b) work on (think about or write) different parts of the paper and then order them.
33. When I have to work on a group project, I first want to
    a) have "group brainstorming" where everyone contributes ideas.
    b) brainstorm individually and then come together as group to compare ideas.
34. I consider it higher praise to call someone
    a) sensible.
    b) imaginative.
35. When I meet people at a party, I am more likely to remember
    a) what they looked like.
    b) what they said about themselves.
36. When I am learning a new subject, I prefer to
    a) stay focused on that subject, learning as much about it as I can.
    b) try to make connections between that subject and related subjects.
37. I am more likely to be considered
    a) outgoing.
    b) reserved.
38. I prefer courses that emphasise
    a) concrete material (facts, data).
    b) abstract material (concepts, theories).
39. For entertainment, I would rather
    a) watch television.
    b) read a book.
40. Some teachers start their lectures with an outline of what they will cover. Such outlines are
    a) somewhat helpful to me.

b) very helpful to me.
41. The idea of doing homework in groups, with one grade for the entire group,
    a) appeals to me.
    b) does not appeal to me.
42. When I am doing long calculations,
    a) I tend to repeat all my steps and check my work carefully.
    b) I find checking my work tiresome and have to force myself to do it.
43. I tend to picture places I have been
    a) easily and fairly accurately.
    b) with difficulty and without much detail.
44. When solving problems in a group, I would be more likely to
    a) think of the steps in the solution process.
    b) think of possible consequences or applications of the solution in a wide range of areas.

In marking your test:
Questions 1,5,9 etc. are Active-Reflexive.
Questions 2,6,10 etc. are Sensing-Intuitive
Questions 3,7,11, etc. are Visual-Verbal
Questions 4,8,12 etc. are Sequential-Global

Appendix 2
Music on the Website

Example 1: The Ill-fated "J L Bach"!
Example 2: The Scale on the White Notes
Example 3: The Scale on the White and Black Notes (Chromatic).
Example 4: The Pentatonic Scale
Example 5: The White Note Bubble Sort
Example 6: The Chromatic Scale Bubble Sort Using the Black and White Notes)
Example 7: Two different Notes. What is the difference?
Example 8: An example of Coordinate (8,4) being auralised
Example 9: Locating a point in the space using Piano and Harp
Example 10: An example of a Line
Example 11: Major and Minor Chord Difference
Example 12: A Successful IF Statement. (Listen for Major Chord).
Example 13: An Unsuccessful IF Statement (Listen for Minor Chord)
Example 14: A CASE statement
Example 15: Examples of the Presentations